A MEASURABLE DIFFERENCE

The Reminiscences of Walker Lee Cisler

In Collaboration with
James P. McCormick

The Entrepreneurial Autobiographical Series
Graduate School of Business Administration
The University of Michigan
Ann Arbor

TABLE OF CONTENTS

DEDICATION

To my friends around the world who have urged me to write this book. For many of them I have become a last voice to help record their accomplishments.

PREFACE

If this stimulating autobiography by Walker Cisler engenders a spirit of enterprise in the young men and women of today on whom the future of our society depends, the primary objective of The University of Michigan Graduate School of Business Administration will have been realized.

The greatness of this nation is tied directly to the achievements of individuals who have served their fellow man well by rendering a most valuable economic and social service in their role as entrepreneur. To recognize such performance is a second objective of this entrepreneurial autobiographical series.

Mr. Cisler has had a rich and varied experience in business extending over half a century. He is a remarkable man. From his simple beginnings as a farm boy, a period in his life he thoroughly enjoys reflecting on, he became a professional engineer largely by virtue of his own drive, and then moved steadily upward through government and private utility positions to the Chairmanship of the Board of The Detroit Edison Company. Today, at the height of his career, he is the undisputed world leader of the energy industry. No man in history has been as active in promoting the welfare of people in many different lands through the development of energy resources as Walker Cisler. We hope this autobiography will be read and enjoyed by a large audience of students and members of the academic community, close friends and business associates, and citizens of the world.

Floyd A. Bond, Dean

Ann Arbor, Michigan
January, 1976

Introduction

This is the story of the business and professional life of Walker Lee Cisler and of his view of the world. With allowances for transferring the spoken to the written word, it is set down as nearly as possible as he recounted it. I have checked data and added background information to provide context, but Walker Cisler has almost total recall, so the narrative comes largely from his memory. In discussing the people and activities that have kept him busy for more than fifty years, he hoped his autobiography would illustrate some of the constructive roles an American businessman can play.

Cisler was born October 8, 1897, in Marietta, Ohio, and after growing up in Gradyville, Pennsylvania, he attended Cornell University. Following a year in military service in World War I, he graduated in 1922 with a degree in mechanical engineering. For the next twenty-one years he was a practicing engineer employed by Public Service Electric and Gas Company of New Jersey, where he advanced to assistant general manager of the Electric Division.

The turning point in his career came in 1941 when he was loaned by Public Service to the Office of Production Management in Washington to help mobilize industry for World War II. Two years later he accepted a position as chief engineer of power plants with The Detroit Edison Company, but almost immediately he was asked to review the utility needs in the Mediterranean Theater. He served as a colonel in the Normandy invasion and in the Allied Military Government in Germany until the end of the war.

The remainder of his career was spent in Detroit. He became president of Detroit Edison in 1951, chairman in 1964 and retired from the company in 1975. In 1972 he organized a small group of retired executives into Overseas Advisory Associates Incorporated, a not-for-profit company that provides consulting services on energy to several countries in the Middle and Far East.

As Cisler told the story, he started not with the chronological beginning but with recent events and the critical years of World War II. A great deal more information about The Detroit Edison Company might have been included, but he was aware that Dr. Raymond C. Miller had already published a two-volume history under the titles *Kilowatts at Work* (1957) and *The Force of Energy* (1971).

Any person who has been so deeply involved in so many activities clearly has great drive and determination. Cisler has learned how to use the resources available to him, and he has used them primarily for others. He moves as easily with ordinary people as with leaders, and he never forgets his rural beginnings. It is only with the greatest difficulty that he can bring himself to say an unkind or derogatory thing about anyone, and where a biographer might make a critical evaluation, he will not. Instead, he is moved by the dignity of men and women that he has found everywhere.

Writing from Belgium, his long-time friend Louis de Heem described him as the best ambassador the United States ever sent to Europe. A selected list of the honors Cisler has received is included in an appended chronology of his life; it speaks eloquently of other people's appreciation of his work in their behalf.

James P. McCormick
Detroit, Michigan

CHAPTER I

WORLD ENERGY

Because I am by training an engineer and by practice a businessman in the electric power industry, I believe in and trust careful design and building of systems. But in October, 1971, an airplane crashed and killed four men with whom I had been associated for many years. It was a personal loss that filled me with sadness and one which I have difficulty accepting.

I had just completed a meeting in New York City that involved financing part of the $2.5 billion expansion program that The Detroit Edison Company was planning in order to meet the growing demands of its service area in southeastern Michigan. The Business Council was meeting in Hot Springs, Virginia, and I joined a number of other executives to advise with representatives of the President of the United States on economic conditions.

Dr. Yoshimichi Hori, director of the Central Research Institute of the Electric Power Industry of Japan, phoned to say he had come to Detroit on his way home to Tokyo and it was important that we get together. He had been in London as chairman of a committee of the World Energy Conference that was planning to participate in the Conference on the Human Environment to be held in Stockholm under the auspices of the United Nations. Since he had never visited an American hot spring, he said he would like to come. I arranged for two pilots to fly with him and three of my associates the next day, and about noon I borrowed a car to drive to the local airport. It was drizzling and the tops of the nearby mountains were cloud covered. Several planes were taking off, but the airport manager had no communication with my friends.

All of a sudden I was startled. A plane came across the crest of the mountains at a sixty degree angle to the runway and passed overhead at about 1,000 feet elevation. The manager thought it was the Detroit plane, but there was still no communication.

About ten minutes later the plane came in again, this time at a thirty degree angle and at about 500 feet in height. We watched it go down the field, and instead of turning left into clear weather, it turned right and flew into the clouds. The manager said the pilot was following the book of instructions and would be back in about eight minutes since he was under control of the Washington D.C. tower.

It took a little longer than eight minutes, however, and my apprehension mounted. As the minutes passed and the visibility lessened, I became frantic in my mind. Then I picked up the sound in the distance in line with the direction the plane should come, but it seemed low because the sound was coming up to me. I was beside myself and started to run into the office to tell the manager to divert the pilot to Roanoke Airport when suddenly there was a silence. It was more than just a throttling back of the engines, more as if the propellers had been feathered, and then there was a sound of a terrific struggle as if the pilot were trying to gain control. Immediately there was a splintering noise as the plane touched the tree tops, and this was followed by a terrifying boom as it crashed into the mountain seventy feet below the crest.

"My God!" I shouted to the manager. "They've all been killed!"

I could see the smoke rising and jumped into my car to drive to the fire. The manager followed in his car and pulled me back from the plane. I was desperate. They were burning up in front of my eyes. I heard the manager speaking and explaining that he thought the fuel tanks would explode. Then other men arrived with the foam truck. Later as they brought the bodies out they said: "Don't look at this." They could find only three who had been thrown forward into the cockpit with the pilot and co-pilot, and I told them there were four and they should look outside. Somehow I knew it was Beekman and that he had been hurled out of the plane.

I think all aboard were killed on impact. We never had a satisfactory explanation as to whether the accident was due to pilot error or to a faulty altimeter or to a combination of circumstances. I felt the crew and passengers were in distress, however, and after the boom of the crashing plane I sensed all those fellows around me in the mist. It was an ethereal feeling, but they came to me and were around me in the environment.

Although my friends were dead I knew that the results of their lives were not wiped out. Each in his own way had added to my appreciation of the world, each had made a contribution to his other associates, and each had left a legacy to mankind. The two Americans had helped engineer, build and operate the largest fast breeder nuclear reactor that had been developed. Enrico Fermi Power Plant #1, named after the pioneer Italian physicist who had started the first controlled nuclear chain reaction, was a demonstration project designed to gain experience and train people for the large commercial models that are expected in the 1980s. Myron Beekman had been with the project from the beginning in the early 1950s and was general manager of the corporation that owned the reactor. Charles Branyan was a nuclear physicist who had joined the project after spending some time with the small experimental reactor built by the Atomic Energy Commission. Kozo Odajima of Tokyo Electric Power Company was completing three years of study of the engineering and economics of the fast breeder before returning to an important position in Japan.

Taken together, we five men represented more than seventy years of effort directed toward producing abundant electricity from atomic fission. I had known men who died in the anger and violence of war. But these men died in the thoughtful and planned pursuit of means that would allow men and women to live with greater understanding of the world, with economic abundance and with an opportunity for peace. Energy makes a measurable difference in the world, just as it has in my life.

The airplane crash led me to recall once again my own career, as I, a farm boy in Pennsylvania, grew up to join the company of men who advise Presidents and other leaders on energy matters. It is a common enough experience in the United States. Many of the businessmen at Hot Springs had travelled a similar road. So I began by asking how all this happened and how I have been fortunate enough to take part in the development of the most varied, the most productive and the most powerful society in history.

Even though I have participated in this growth, I never forget that people have not always enjoyed abundance and that even today there are millions who suffer great privation. An adequate supply of energy makes the difference between hardship and comfort, between primitive existence and civilization.

At some early period before recorded history, men and women began to use fire, and this discovery made a major change in their lives by adding the heat of combustion to the small supply of energy they found in food. They lived by hunting animals and other wild foods which yielded the 2000 calories of heat energy required in their daily diet for sustenance. And they could keep warm with the fire. This condition lasted tens of thousands of years until another inventive and peculiarly human step was taken with the domestication of animals and grains. Enough secure energy was then available in the form of food and muscle power to permit population densities to rise and permanent cities to replace transient campsites. But it was only about 4,000 years ago that these first city cultures were established in river valleys where the climate and soil conditions were especially favorable for agriculture.

It is estimated that these neolithic farmers had about eight times more energy available to them in the form of food and domestic animals than the paleolithic hunters. With the introduction of copper and iron to replace the stone instruments, simple machines like the plow, lever and wheel were invented which improved the productivity of work. The sail was a major technological discovery because it joined the force of the wind with the buoyancy of water. One of the early, highly organized, industrial systems of men and machines was found in the sailing ship. Again it is estimated that the people of the high Roman and Renaissance cultures had about thirteen times the energy of primitive hunters, and this was enough to free a few of them to realize the tremendous potential of the mind and spirit.

As a boy I grew up in an environment that was closer in many ways to the early agricultural life than it was to the world of my present daily activities. I always remember this. It has been a source of satisfaction to me to have lived in a period and place when, for the first time, whole nations have begun to have enough energy to satisfy their basic needs for food, shelter and transportation. Americans today enjoy forty times the potential productivity of primitive man. It is forecast they could have 150 times as much by the year 2000, and this amount will indeed be necessary if we are to extend a high standard of living to all of our citizens while doing our share in improving the quality of life in the rest of the world.

4

The World Energy Conference

In many ways the fiftieth anniversary meeting of the World Energy Conference held in Detroit in September, 1974, represented a summation of my professional life. I had been in the industry for fifty-two years, especially in the production and distribution of electric power, so I had lived through much of the developing industrialization of people around the world. I had been president and then chairman of The Detroit Edison Company for twenty-four years and had taken part in energy studies in twenty countries at the request of the State Department. And I had been chairman of the International Executive Council of the World Energy Conference from the occasion of the Moscow meeting in 1968 to the gathering in Detroit of the national committees from seventy-one member nations.

Upon graduation from Cornell University in 1922, I started to work in the power industry, where I was employed by Public Service Electric and Gas Company of New Jersey. Leaders of the industry in the United States were just beginning to talk with their colleagues in other countries following the disruptions experienced in World War I. Daniel N. Dunlop, director of the British Electrical and Allied Manufacturers' Association, obtained the financial backing of his board to organize an international assemblage of scientists and engineers who were interested in energy, especially in electric power. A thousand participants came to London in 1924 from about thirty countries to look at the constructive uses of energy for peacetime. I remember reading about it in the newspapers and technical journals because some 300 sailed from the United States in one of the largest delegations of its kind. The theme of the conference was "Resources of the World in Power and Fuel and Their Use to the Greatest Possible Advantage." So successful was this first gathering that it was agreed to organize the World Power Conference.

Each of the national committees had up to four delegates but only a single vote on the International Executive Council, which met at least annually. The council planned the plenary session of the conference every six years to consider broad energy questions and a varying number of sectional meetings on more limited topics in between. The president of the conference has always been chosen from the nation that was host of the latest plenary session, but the office of the secretary general has always been located in London. From the beginning the members have thought of it as a worldwide organization for sharing technical ideas and not as an adversarial body to advance the economic interests of a particular country. This minimized political controversy and encouraged the interchange of information that, like electricity itself, could flow across the boundaries of local company franchise or national territory to the person who pushed the switch because he or she needed the power for some work at hand.

At the time of Dunlop's death in 1955, it was noted in the memorial remarks that he saw the conference as a "meeting place between scientists and engineers on the one hand, statesmen and economists on the other." It is my own observation that his vision of the organization as a means of bringing concerned individuals together has proved as valuable as the technical information that is exchanged.

Following World War I the use of electricity in industrial countries began to double every seven years, and investor-owned power companies were organized to meet this demand. Whereas, just a decade earlier it was still a novelty to enjoy electric lights and appliances, when I began to practice engineering about 10 percent of all the energy used in the United States was in the form of electricity. The worldwide depression of the 1930s halted this growth, and many people were forced to return to kerosene lamps to light their homes. About 1935, when the use began to increase again, economists noted how closely this related to the growth of output as measured by the Gross National Product.

The Third World Power Conference was held in Washington in 1936, at the height of the New Deal, when Secretary of the Interior Harold L. Ickes was a strong advocate of public ownership and nationalization of the power industry. In the Soviet Union, of course, the industry was completely nationalized, and many of the European governments were moving in this direction. I read about such things as the World Power Conference, and I was aware of the political differences between the advocates of public and private development. But my job was to find answers to technical and economic realities within the State of New Jersey, and the meetings of the Conference were far above my position.

With the outbreak of World War II, the activities of the World Power Conference were laid aside along with the move to nationalize the power industry in the United States. Electricity had become an essential part of our daily lives. In 1945, I was stationed in Berlin as chief of the Public Utilities Section, Office of Military Government of Germany, and Sir Harold Hartley sought my help. He had succeeded Dunlop as chairman of the International Executive Council and wanted to direct responsible people toward the reconstruction of Europe as quickly as possible. This was my first direct contact with the organization, and I liked Hartley immediately. As a natural science tutor at Oxford University who had developed the Balliol College laboratory in physical chemistry, a member of the board of the Gas Light and Coke Company, a later vice president and director of research of the London, Midland and Scottish Railway, he had an unusually broad background. He was also active in the development of the British airways, another energy-intensive industry. Because of my position, I was able to help get representatives from the allied countries to plan the World Power Conference to be held in London in 1950.

By that date, however, the Marshall Plan was in operation, the Organization for European Economic Cooperation had been formed, and I was very active in the rebuilding of the European power systems. I also had my principal job at The Detroit Edison Company, so I was unable to attend the conference. Over the next twenty years, nationalization of the power industry continued to grow in many parts of the world, but fortunately there was a respite from this pressure in the United States which helped us to maintain our position as the leading industrial nation.

During the 1960s I became a member of the U.S. National Committee, and when Lord Hinton of Bankside, who was then chairman of the International Executive Council, set up a review committee, he asked me to serve on it. By this time I had participated in energy and economic studies in many parts of the world and was convinced that far broader considerations than electricity alone were required. In addition to worldwide membership, the conference needed comprehensive programs that included all energy sources. If nations were to have adequate supplies in the future, they would need to use water power, fossil and nuclear fuels that were already available and to start developing even newer sources.

The charter and bylaws of the organization were then forty years old, and I proposed they be reexamined. This process took three years of discussion. In 1966, at the regional meeting in Tokyo, I invited the conference to come to Detroit in 1974 because I had come to feel increasingly that we needed to make the transition to the larger objectives. The invitation was received enthusiastically and they said, "Cisler, we'll make you chairman." When I protested that I would be fully occupied arranging matters in Detroit, they said jokingly, "We'll take care of you." And they did. In Ghana the following year I was named chairman designate and at the Moscow Conference in 1968 became the first person outside of England to be elected chairman of the International Executive Council.

The review committee recommended that the name World Power Conference be changed to World Energy Conference to remove any association with political power, or even exclusively with electric power, and to emphasize the total field of energy. The committee also decided that a statement should be added to the objectives of the conference that would stress the relationship of energy to economic growth. Finally, it recommended that the sectional meetings be eliminated and the worldwide significance of energy be recognized by holding the plenary sessions every three years. By this time, there were sixty-five member nations that needed to exchange information on population increases, increasingly scarce fuel resources and technological improvements for greater efficiencies.

These changes were approved by the executive council in Moscow in 1968, and the conference that year was held around the theme: "World Energy Resources and Their Utilization for the Welfare of Mankind." In addition, Pyotr Neporozhny, minister of power and electrification of the Soviet Union, was elected president of the conference for the next three years. He had just been awarded the Lenin Prize for the design and construction of the largest hydroelectric plant in the world near Lake Baikal.

In Moscow, we learned again that the World Energy Conference existed within a political-military environment, and that efforts were needed to preserve forums which pursue rational solutions rather than force. I had met with Neporozhny early in the day before the official opening, and when I returned to the hotel the American embassy called to report the Russian invasion of Czechoslovakia. Almost 6,000 delegates and guests were gathering at the Palace of the Congresses in the Kremlin, and the news spread quickly. The conference was thrown into disarray.

I went directly to Neporozhny and said we must help stabilize the situation and not let it upset the conference. Then Czechs came to ask for help in getting home because there were almost a hundred members of their delegation who thought they might be interned. Since I was chairman designate, people wished to see what I was going to do and how the American delegation would respond. The situation became very tense, and a few of the national committee representatives left while the rest of us kept everything in as low a key as possible. I talked with the other national committees, and many of them exerted a calming influence. They assessed the situation as a matter between the U.S.S.R. and Czechoslovakia that was not going to develop into a general conflict. Finally, I was able to report back to the American embassy that we were going to stay and carry on the conference.

Under these trying circumstances Neporozhny was just wonderful. He got two planes to fly the Czechs back to Prague. The Russian delegation was host to the world and was distressed by the events that were being reported but which were so clearly beyond their control. Our hosts assured us they had been unaware of the situation and, in fact, there was no visible military activity in Moscow. So they went about the business of the conference and did not talk about conflict or ideological differences.

The national committee of the Soviet Union acted responsibly and with deep understanding of the anxieties that had been created. I said to Neporozhny: "What you do in your country is your affair. The other delegations feel this is an internal matter. But it should not prevent us from doing good things together. You have the same instincts I have. Nothing will happen in the right way if our concern for other people is not right, because this is fundamental."

In the final analysis, it is the human objectives and judgments that are displayed which are important. If an individual is not honest with himself and others, one cannot work with him. The real test of action in any situation is: What is fair to the people who are involved? Neporozhny and the Russian delegation were concerned about others. They did everything in their power to quiet fears, and the conference was a success. The 270 technical papers and eleven studies of the broad aspects of energy economics had been printed and circulated in advance according to the usual practice. The facts of the invasion were on everyone's mind, however, and the value of having private businessmen and government officials, economists and engineers, politicians and scientists meeting face to face was again established. Individually, everyone was shaken, but the conference survived to bring people together on other growing pressure points.

I immediately turned my attention to my term of office. The following year at the Executive Council in Ankara, Turkey, I talked about the imbalance of people, natural resources and energy, saying: "We ourselves and our associates in the energy field can encourage and show the way toward economic progress and international understanding because energy and power provide a basis that all men can understand and benefit by. These are universal in character." The eighth World Energy Conference in Bucharest, Romania, was held without incident and was largely technical in nature. But the forces of supply and demand were gathering speed in their headlong flight in opposite directions, while political manipulation was producing contradictory government policies that spread confusion.

It was in 1974 that the critical forces of the quantity, distribution and need for fuel in all its existing physical forms collided with one another and forced a readjustment of existing economic systems. For the first time since World War II, demand was so much greater than supply that serious dislocations were produced in the industrialized nations. This spread instantly to the agricultural nations, which in earlier periods of history would have been isolated from the effects. But 1974 proved disastrous to them also. The cost of energy, which had been moving lower for two decades—and indeed for two centuries—suddenly reversed direction and moved sharply higher. The petroleum exporting countries quadrupled the price of oil. The resulting foreign trade deficits and lowered productivity created shortages of food and shelter and other elementary human needs on a scale never before experienced in so short a time.

By 1974 it was also clear that the industrial world, which had been created by the work of men and women, was having a serious impact on the natural world we had inherited. As people looked at spaceship earth for the first time from the vantage point of television cameras on the moon, the desirability of protecting our natural resources grew. Assumptions about our environmental birthright were strained by a population explosion, and we suddenly realized that relations among human beings and the air, water, soil and other living things were being endangered.

The World Energy Conference, meeting in plenary session in Detroit, celebrated the fiftieth birthday of the organization in an atmosphere less explosive than the Moscow Conference but far more critical in worldwide implications. The theme, "The Economic and Environmental Challenges of Future Energy Requirements," was chosen to focus the attention of responsible policy makers on the divisive issues. It resulted in much greater clarity of thinking and in facing up to some of the painful alternatives among the seventy-one member nations and the nine others that sent observers.

It has been customary for the conference to be opened by an important official of the host country. The Prince of Wales addressed the London meeting of 1924. The Detroit meeting of September, 1974, was opened by President Gerald R. Ford, speaking from the most powerful political position in the world.

In the spring of the year, we had invited President Richard M. Nixon to open the conference. Spiro Agnew had resigned as Vice President, and Gerald Ford had been nominated to succeed him, so we had approached Mr. Ford to be our closing speaker. Meanwhile, a number of the Arab states had agreed to send representatives, and as late as July, I was trying to bring the Cubans but failed to get policy changes in Washington. Then, in August, Nixon resigned, and President Ford, who had been a member of the House of Representatives for Michigan during a quarter of a century, agreed to speak at the opening.

Governor William G. Milliken and I met Ford at the airport and drove with him into the city. On the way he asked me to talk about the energy situation and I told him about the imbalances existing around the world in relation to population, natural resources and systems of productivity. I described how the Marshall Plan had been organized after World War II to meet a similar crisis and how Michigan was the one state that had made its own energy survey. He invited me to meet with him after the conference.

President Ford opened his remarks by saying: "The whole structure of our world society rests upon the expectation of abundant fuel at reasonable prices," and since the "expectation of an assured supply of energy has now been challenged," the "repercussions are being felt worldwide." Then, after noting that in the past nations have gone to war in such situations, he said: "American foreign policy rests on two obvious new facts. First, in the nuclear age, there is no rational alternative to international cooperation. Second, the more the world progresses, the more the world modernizes, the more nations need each other."

Ford used the platform to announce a shift from the Nixon goal of Project Independence in fuel to an achievable and non-threatening policy of Project Interdependence with the whole world. With this important pledge toward international cooperation to ease the tension between the supplying and consuming nations, he then described the "disastrous consequences if nations refuse to share nature's gifts for the benefit of all mankind." If effect he was asking the petroleum exporting countries to moderate the aggressive position they had taken by quadrupling the price of oil.

The President of the United States was answered by Sheikh Ahmad Zaki Yamani, minister of oil of Saudi Arabia. He said the position of OPEC was justified.

He said it politely and in the language of economics, but he said it unmistakably. He noted that the oversupply situation existing in the 1950s and 1960s had been reversed by a ten-year average increase of 7.5 percent a year in consumption. As a result, the long-term price of about $1.80 a barrel had increased rapidly, and this was creating serious balances of payment deficits for the industrial nations. His essential answer to President Ford was that a solution to the financial problem was "largely contingent upon the cooperation of both developed and developing nations with the oil producing countries in creating a healthy environment for the investment of surplus funds."

The answer continued to be adverse. Canada, Iran, Venezuela and the other oil exporting countries have moved to conserve their resources. The United States can expand domestic supplies most quickly by increasing the use of coal, of which this country has ten times more than any other form of fuel. Yamani pointed out that coal from the western states could be delivered in Detroit for the equivalent of oil at $5 a barrel and eastern coal for $4 a barrel. This was in contrast to $12 a barrel oil from the Middle East. Nonetheless, the environmental impact of this change would be significant and the dollar costs much higher than Detroit Edison customers had been accustomed to paying. Europe and Japan did not have even this expensive alternative, so they left the conference determined to increase their nuclear generation as rapidly as possible. Despite such new moves there will be a serious decline in real output and in the standard of living until the energy balances are restored.

Four Imbalances

From my experience I have learned that the world as a whole contains ample resources to enhance both the economic advantages of people and the physical environment. What is needed is continuing inventiveness and better management. What is lacking is lead time. In 1945, The Detroit Edison Company could provide an adequate response to increased demand from its customers in two years. In 1975, it had to plan ahead at least twelve years to be reasonably confident it could meet the needs of its service area.

The four categories in which I have seen nations contending with imbalances are population, natural resources, energy and technology which includes educated people to make it productive.

The first imbalance concerns the number and distribution of people in the world. At the beginning of the industrial revolution about two centuries ago, there were approximately 700 million people, and by 1860 the estimated world population was 1,000 million. By 1930 it had doubled, and by 1960 the increasing growth rate had raised it to three billion. By some estimates there could be as many as seven billion people trying to utilize the resources of the earth by the year 2000. Today, more than half the population lives west of the Philippines in China, Southeast Asia, India and Pakistan. And it is in these developing countries where the largest increase is appearing.

The second imbalance is found in the distribution of such natural resources as weather, water, land, living things and metals. With fertile topsoil and a temperate climate there are land masses that can sustain large populations. Many parts of the world, however, are natural deserts like the Sahara with inadequate rainfall or like the Canadian and Siberian tundra with cold temperatures. Others, like parts of the Mediterranean basin, have been overgrazed and overworked for centuries. Minerals also are poorly distributed.

The third imbalance is the natural distribution of energy resources. The earth is amply endowed in some locations and poorly in others. Some of the small population of Arabia live over great reservoirs of oil. There are reliable estimates that the coal reserves of the United States are large enough to supply this country for 500 years whereas Japan has almost none. In Greece there was only low energy lignite available to start the postwar recovery program. In India a very large percentage of the fuel for cooking and heating is supplied by the dried manure of cows and camels.

Man-made shortages, however, are more important than natural ones. When I began my career in the 1920s, coal supplied almost 85 percent of the primary fuel converted to heat in the United States, but today petroleum products have become the dominant energy form. Since these changes have resulted as much from increased human demand as from decreased natural supply, they have been managed with remarkable success over this half-century. When stable political and economic systems break down, however, the control of fuel can disappear with a suddenness that affects people seriously.

The fourth imbalance lies in the amount and distribution of technology. Industry, which is a rational organization of men and machines, produces a high standard of living for a few nations and an increasing gap between them and the developing nations. The basic elements of this situation are quite simple. Throughout most of recorded history, wealth has been in animals and land. Since I grew up on a farm, I knew these values at first hand, although it was not until I was an adult that I began to think about an agricultural economy. It was the invention of the steam and the internal combustion engines, however, that really extended man's muscle since they could burn unlimited amounts of energy. Because machines were so immensely productive, man was freed from the necessities of physical labor. I, myself, was able to lay aside four years for study at Cornell University and to become a manager of systems of machines and the men who make them perform. But along with the shortages of machinery in the developing countries, there are inadequate levels of education and the resulting management skills. Only increasing productivity provides the extra margin to bring people, resources and energy together for social progress.

Despite the vast amount of water power and fossil fuels people use today, we need much more if the standard of living is to continue to improve. A good deal of energy can be conserved by designing more efficient machines and industrial systems. The World Energy Conference meeting in Detroit made arrangements to set up a conservation commission. But mankind appears to be profligate by nature, and I have met few people in my lifetime who have not found pleasure in some form of lavishness, if only in the giving of gifts to children. If they have a choice, most people prefer abundance to scarcity.

Fortunately, the discovery of a new source of primary energy in the atom has come at just the right time. Scientists have already gained the necessary understanding of how to sustain the fission of the uranium atom in a chain reaction. Engineers have built power plants to turn the resulting heat into electricity. One of the things that impressed me about the Detroit conference was the determination of the delegates to return to their homelands and double their nuclear generating capacity. There are forecasts that up to 40 percent of all electricity will be produced in nuclear plants by the year 2000. This is one of the great quantum leaps in human progress, and it can supplement other fuels for centuries.

Scientists are presently investigating the nature of fusion of atoms of hydrogen, deuterium and tritium, all found in quantity in the waters of the oceans. Whereas fission uses an explosion to split the elements of an atom of uranium, fusion uses an implosion to drive atoms together in a thermonuclear reaction. If engineers can build machines to utilize the resulting heat on a large commercial scale, there will be a limitless supply of energy to meet the needs of people everywhere. Under the most optimistic timetables this is a development for the next century, and in the meantime we must fill the shortage of fuel with uranium.

With the discovery of fission and fusion, mankind has entered upon a quite new era. Throughout history we have been dependent upon the process of combustion; we created heat by burning wood or coal or oil. By learning how to control the new processes, mankind can finally duplicate on earth the fundamental energy cycles of the sun itself. No other age has had so much natural resources, technology and management experience with which to work for the benefit of all mankind. I believe deeply that energy and human development are closely interrelated. Neither can be advanced without the other. For the first time in history there is a real possibility of eliminating hunger and poverty from every region of the world, and such a goal is worthy of the best that is in us.

All these matters were subject to searching analysis during the World Energy Conference in Detroit. In his summary of the many discussions, Wilson W. Campbell said: "We have met once again as practical people, but in an atmosphere more than ever economic, cultural, political and philosophic. It's no bad thing that a forced transformation in energy balance should drive us back to philosophic principles."

I, myself, feel that the conference has made the transition to an organization that will be as helpful in the years ahead as the old conference was during its half-century. While I would have much preferred that the transformation of energy economics for the next quarter century had evolved voluntarily rather than been forced suddenly upon the peoples of the world, the World Energy Conference will continue to bring together individuals of knowledge, judgment and goodwill to help solve the many problems in rational ways.

Upon completion of my major assignments another question has been raised. What does a man who is seventy-eight years old do with the rest of his life? The answer seems to be: Many of the same things he did in the early part except on a reduced scale. I remain active in voluntary and philanthropic work. And, along with some other retired executives from the electric power industry, I have organized Overseas Advisory Associates Incorporated.

Officials in Washington continued to make sometimes urgent requests for assistance in putting teams together to go into the field and deal with matters at first hand. My activities had kept me continually involved with people in several parts of the world since the days of lend-lease, so I could be helpful. In discussing the need for systematic steps to build well-designed energy organizations, I realized it was desirable to have a small group of veterans who were free to work on some of these concerns. We did not want to compete with large companies on long-range projects, so we organized a not-for-profit corporation with a roster of individuals who were willing to let others cut the grass while they were away. Sometimes only one man was involved although usually some three to eight people formed a team.

Our first main undertaking was in Vietnam to assist in the development of the Vietnam Power Company. We then resumed some earlier work I had done in Iran, particularly with Teheran Regional Electric Power Company. This was followed by a study in Bangladesh and most recently in Saudi Arabia. Through ventures such as these we are still able to do useful work.

CHAPTER II

THE WAR YEARS

My reminiscences as a businessman really begin in 1941, when I was forty-three years old, because it was during the years of World War II that I first began to understand how business and government could work together and how to involve one nation with another. Before this, I had served briefly in World War I as a lieutenant in the field artillery, had graduated from Cornell University engineering school, and had been employed in a middle management position in Public Service Electric and Gas Company of New Jersey. It was a large company, and I experienced the rapid growth of the 1920s and the depression of the 1930s as a professional engineer dealing with very real problems.

In World War II, however, I worked with many big corporations all over the United States and in Europe, and I dealt with top management. By happenstance, I was one of three people made available by the power industry to go to Washington as consultants during the period of lending and leasing essential supplies before the United States actually entered the war. The needs of our allies had imposed serious strains on the American economy, and the Office of Production Management (OPM) was organized to provide priorities for civilian and military needs. William S. Knudsen, president of General Motors, was director general, and Sidney Hillman, president of the Amalgamated Clothing Workers, was associate director general with special responsibility for the labor division. These were two of the largest organizations in the private sector, but size was a relative term, and the long-standing concerns about big business, big unions, big government and big military simply disappeared. None of them was big enough or well enough organized to deal with the war. The dispute over private vs. public power development was laid aside. The additional requirements of both lend-lease and our own military buildup were throwing the industrial system out of balance, and action had to be taken to restore it.

I was appointed to the Power Division of OPM on September 25, 1941, to help break the log jam in priorities that had started to pile up.

The division was responsible for electricity, gas, sewer and water. The last three areas presented relatively few problems, but the demand for electric power expanded very rapidly. The manufacturers already carried orders on their books for about 125 turbine generators of 10,000 kw and over, and about 200 for units in the 500-7500 kw range. This was a two-year backlog at full production assuming no shortages of scarce supplies. And in September, I represented the entire Manufacturers Unit of the Priorities Section in the Power Division of OPM. Many people, perhaps all of us, were rather naive in those first days about the size of the task. My assignment was for three months, but I did not leave government and military service until four years and many adventures later.

In terms of my experience, the war was the greatest challenge to American individualism and the free enterprise system in the twentieth century. People went into it under conditions of legal, political and social constraints that had never been felt before with such force. The free market was subordinated to wartime planning. Profit was subordinated to resource allocation, as gasoline and food were rationed. Economic growth was subordinated to victory, as wages were fixed and ceilings placed on prices. Individual rights were subordinated to military conscription, as production people were frozen into their jobs. The question before all of us at the time was whether the American people as a whole could mobilize. The question before industry was its relation to government and the armed forces.

War Production

The Office of Production Management was known familiarly as Oh Promise Me because it had established a priority system. The trouble was that no one understood exactly how the production of really critical items was affected by these overriding priorities, and it was hard to deliver on the promises. The demands for heavy power equipment, ships, aircraft, tanks, gasoline, synthetic rubber were all competing for scarce resources, and it was not good enough to say that the military requirements had first priority, public interests represented by the federal government had second, and the civilian economy had third. They were all interrelated and dependent on one another.

I started immediately to travel around the country to inspect the manufacturing facilities and to talk with managers about their operating problems. The Power Division under Julius A. Krug, who had been brought to Washington from his position as manager of power for the Tennessee Valley Authority (TVA), began to make studies of the amount of electricity that would be needed for both civilian and military needs. Each utility company reported its estimated load requirements and its generating capabilities. The geographical regions and power areas, as defined by the Federal Power Commission, were used substantially without change. This information permitted us to meet critical needs as they developed and to locate new industry. where the supply of electricity was adequate. Peak loads for 1942-44 were predicted accurately many months in advance, and the industry was able to meet them.

As a result of my inspection tour, I realized that the military services were making indiscriminate use of their overriding priorities and that the country would soon run short of electric energy. I had seen a Navy captain standing beside the production line of General Electric in Schenectady and bumping the generators that were being built for domestic installation. I then attended a day-long session called by Admiral Samuel M. Robinson who was chief, Office of Procurement and Material. He started from the position that the fleet needed generators both for the ships and for bases around the world. During the day his group outlined what General Electric, Westinghouse, Allis-Chalmers and others could produce. They were concerned that a forging for a turbine generator was in competition with a forging for a ship, that steel for a boiler was not available for armor plate. I listened to all this, and about five o'clock I raised my hand and said: "Admiral, may I have a word? I have listened attentively all day, and you have scheduled the total manufacturing capacity of these companies and left nothing over for civilian production. You are not going to have enough power to produce the equipment you say you will need."

This was a challenge to him, and he came back at me and said: "Hell, any blacksmith shop can make the turbines for land use, and you can make them out of clay if you really have to. You can turn them out of ceramics. You don't need steel."

I reported all this to Krug and recommended that we forget about the priority approach and schedule production to meet our overall needs. I then wrote a letter to the manufacturers ordering them to produce turbines for land use and for marine use in a given sequence and to reject anyone who said another priority took precedence. Krug signed the letter late Friday afternoon.

On Monday morning all hell started to break loose. People began to ask, "What are you trying to do?" and to say, "You have no authority." And, indeed there was no legal authority behind it as our general counsel, John Lord O'Brien, informed us. But people accepted the letter because it brought a sensible order out of the priority mess. Everyone wanted to operate intelligently, and this seemed to offer a way to do so.

Shortly afterwards production began to fall into line, and this success, along with others in steels and chemicals, led to what was known as the Controlled Materials Plan (CMP). It took many hours of intense effort to convince the admirals and generals that certain civilian programs had to go forward in order to help the direct war programs. When they visited the shops, they could see flowing down the production line in orderly sequence a turbine for a battleship, a second for a power plant in a utility system, a third for a troopship, another for a synthetic rubber plant. Much was accomplished not because of the authority of the law, but because people understood the goals and had confidence in one another. It is my belief that this is the greatest force that can be brought to bear on any problem, because if people agree, then barriers are overcome and common objectives obtained. In time, we gained the confidence of the military. We earned it because the power supply was never too little or too late.

We had some experienced people together by this time in the Power Division of the OPM. In addition to Krug, who became secretary of the interior after the War, David E. Lilienthal had a background in the regulatory agencies in Wisconsin and then worked for the TVA. Joseph C. Swidler also had an extensive TVA background. He later became chairman of the Federal Power Commission, and then chairman of the New York State Public Service Commission. The Controlled Materials Plan was the key to productivity. When the synthetic rubber project was approved, resources had to be made available so that it could be carried out. The steel industry supplied people to schedule steel production, and we scheduled the manufacturers for the production of power equipment. All this was coordinated in the War Production Board, which drew the people from industry to perform this great national undertaking. Government people were called upon in turn to do the administrative and legal matters. By March of 1943, it was possible to organize the Power Division into the Office of War Utilities.

Meanwhile, on December 7, 1941, the Japanese navy struck Pearl Harbor, and the country had to unite, mobilize in its totality and rise in its own defense. The Office of Production Management was reorganized into the War Production Board and became the supreme authority over industrial production to meet the war needs and the civilian requirements.

A small project that was underway when I got to Washington illustrates the ingenuity and cooperation of the times. Knudsen approved the production of four floating power plants with a capacity of 30,000 kw each. The purpose was to move them around the inland waters of the United States and tie them up at any point where there was a shortage. When the *Inductance* was completed two years later, she was moved to Vicksburg, Mississippi, and her power was shipped to Arkansas to gin the cotton crop. New aluminum plants in the area were taking most of the local generation, and the barge made up the shortage. Actually this was the only one of the floaters to serve the civilian economy. The others served the Army overseas.

In addition to talking with our own industrial people in the first two months of my assignment, I also met with representatives from Great Britain and the U.S.S.R. on lend-lease requirements and learned of the great destruction in the Ukraine. One-half of the ten million kilowatt capacity in Russia had been destroyed in the blitz. So we scheduled one of the barges for overseas, had a bow welded on and steering gear installed. It was christened *Seapower,* and I proposed locating it in Alexandria, Egypt, so it could be moved to the Black Sea and Sebastopol when North Africa and the eastern Mediterranean were cleared of German operations. This was my first contact with engineers from the U.S.S.R., and it formed the basis of the many years of association I have had since then.

Then our advanced planning indicated that there ought to be a floater for the European theater of operations, so we considered diverting a 25,000 kw generator that had been produced for a copper project in Utah and converting it. But the super-secret atomic project at Oak Ridge also needed additional power for uranium enrichment, so in the end we shifted three units from Utah to Tennessee. Then we worked on new equipment for the floaters and put that design together by working night and day in my quarters in Washington's Dodge hotel. Two Detroit Edison men were in on this.

Frank Nitzberg was in charge of the installation of the generating equipment. Twenty years earlier he and I had been in the same dormitory and the same class at Cornell. Jim Parker of the company and I had attended his wedding in Ithaca after graduation. Subsequently, he had become one of the leading boiler experts in the country and later moved to Detroit.

William W. Brown was an engineer, trained in Scotland, who also worked for Detroit Edison and was now in Washington. He already had a reputation for being able to do more than anyone else with pipes and valves, so he was made responsible for simplifying the designs and producing an efficient, stripped-down version to save scarce material.

As a result of good planning and great effort, three of the barges were completed in 1943. My daughter Jane, who was eight years old, christened the *Resistance*, which was fitted with a bow for the trip to Europe. The fourth barge, the *Impedance*, was completed in 1944.

When General Douglas MacArthur had recaptured the Philippines, he sent a flash to Europe asking for two of the floaters. The *Impedance* was sent ahead through the Panama Canal and provided the first power for the reconstruction of Manila. The *Resistance* followed from Europe. Later, during the Korean War, they were moved to Pusan and Inchon. Later again they were towed to Okinawa, where one of them is still operating, while the other is generating power in Guam.

After serving in Europe, the *Seapower* was sent to Puerto Rico for a tour, then to Rio, and today it is operating in southern Brazil under the name *Piraque,* the Portuguese equivalent of *Electric Eel.*

Until 1943, the fighting did not signify to me a personal disaster to individual men and women. Although I did consult with the Air Corps on the most efficient way to drop bombs in order to knock out power plants, I was still remote from the actual scene of destruction. I had been given some very complicated problems to solve, and I brought my early experiences to bear on the solutions. The war meant new ventures, excitement and an opportunity to perform one's patriotic duty. Then in October a telegram signed "Eisenhower," who was headquartered in Algiers, came to the Adjutant General in Washington. As a result of the fighting in North Africa, the utilities had been destroyed, and now the invasion of Sicily and Italy was creating similar difficulties. The telegram asked for an experienced person to be sent over.

Meanwhile, my original assignment with the Office of Production Management had been extended from three months to two years, and the supplies were now flowing smoothly. My trips around the country had brought me to Detroit several times and, furthermore, I had come in contact with Alfred C. Marshall, the recently elected successor to Alex Dow, who had led Detroit Edison from its beginning almost forty years earlier. James W. Parker, whom I had known from my days at Cornell, was now vice president. I had a good feeling about the company. It was engineering oriented. It was very human and thoughtful of its own people and encouraged their participation in community affairs. In line with its policy of freeing people to serve the public interest, fourteen key men were made available to the War Production Board at a dollar a year while the company paid their salaries. They were working engineers of great talent who knew how to get the job done. No corporation can buy such a reputation for responsibility. It has to be earned. So, when Jim Parker came to Washington and said, "Walker, we really want you to come with us to help us in the postwar period," I accepted. On October 1, 1943, I officially joined the company, although another delay occurred, and I did not actually report to Detroit until a full two years later.

Military Service

General Eisenhower's telegram came to the attention of General George Marshall, who was concerned with the destruction in the operational areas of North Africa and Sicily. It was essential to restore gas, electric and water services as quickly as possible for both the military and civilian activities, a task quite similar to the mission of the War Production Board.

Almost immediately an agreement was reached with The Detroit Edison Company to let me enter the military service for three months so that I could go to the Mediterranean Theater, make recommendations and then be mustered out of the service. Despite the good intentions of everyone, this proved another naive decision. Although I could have had a higher rank, I was commissioned a lieutenant colonel because this seemed adequate to what everyone agreed would be a short mission. I was reissued my World War I serial number, and then made my first air crossing of the Atlantic. And one assignment led immediately to another, so I was involved for the duration.

In World War I, General Pershing had a working government in France as an ally, and he could concentrate on defeating the German Army. But now the allied governments were in exile, and General Eisenhower was faced additionally with the care and welfare of millions of civilians in the countries to be liberated. Before he could decide on D-Day, the actual date of the invasion, he had to assemble experienced technical people, both civilian and military. He had to accumulate the equipment and supplies in England to provide essential services and to keep them flowing, principally from the United States. And he had to organize this tremendous undertaking among many military formations and civilian agencies.

Within a matter of days, I landed at Palermo and spent two weeks surveying the utility resources of Sicily. Here I learned about war at first hand, and here I began to understand what was needed to put the electric, gas and water systems back together again. It was on the island of Sicily, in the midst of destruction, that I vowed to myself that if ever in the postwar period I could use my knowledge to help bring about a more peaceful world, I would do so. Because the need for energy was so common everywhere, it was like a universal language, and I hoped that through it I could talk with men and women about productivity, security, education and a life full of abundance instead of scarcity, fear and ignorance.

I went next to Naples to survey the destruction of the utilities of southern Italy. In January, 1944, I was sent on a mission to the British in Tripoli to examine the power system and rehabilitate the steam generation station there. Finally, at the end of four months in the Mediterranean, I was transferred to the European Theater because Eisenhower had been ordered to prepare for the Normandy invasion. I felt sure there was not enough planning to meet both the military and civilian needs for electric power.

When I left the Mediterranean, I had enough sense to have orders prepared for just two weeks temporary duty in London with the planning organization. The American officers at Norfolk House included veteran utility men, but they had no idea what to expect under wartime conditions or how to deal with the problems. So my photographs of the actual destruction of plants and transmission lines under air and ground attack were valuable. I had also seen the work of putting the pieces back together again, and I reviewed with them the needs for generating equipment, transformers, and other supplies necessary to restore service to cities and military installations.

After two weeks, I completed my briefing duties, and on February 13 I started home with the supply list. I was made a courier so that I could return quickly, and I flew to Prestwick, Scotland to make the run. It was a memorable occasion because I was going home. I even had an extra day, so I drove over to the town of Ayr to visit the birthplace of Robert Burns and to look at the farming country described in his poems.

We took off at night with a crew of seven and reached the Azores in the early morning without incident. The men brought crayons and construction paper along and made valentines for their wives and children. After we refueled, the pilot decided to take the Bermuda route instead of the Newfoundland, and we started again. For the next twelve hours we fought a severe storm. The navigation was very difficult. At times we had to fly low to determine the wind drift from the action of the waves, and the turbulence was severe at all altitudes. As we approached the United States, we received reports that all of the mainland airports were closed because of the weather. This meant that we had to make a landfall on Bermuda Island. We did not have enough fuel to fly around the storm. For a while we thought we would not make it, that we would have to ditch the plane in the ocean. But with just a few minutes flying time left, we finally sighted the island, and the pilot dove directly onto the landing strip. He had flown the ocean many times and had carried people and supplies to the Middle East and India. He was very calm and confident the whole way, which was more than could be said for me. But I got up early the next morning and took one of the traditional bicycle rides before we flew on.

In Washington I reported to General Lucius D. Clay, who had charge of dealing with the civilian population in Europe, only to discover that Eisenhower had sent a wire for me to return. By this time I really had more experience under forward wartime conditions than most from the utility field. Clay told me to visit my wife and children in New Jersey while he approved my equipment list, including *Seapower* and *Resistance.* I checked with Jim Parker in Detroit and received an extension on my leave of absence. Within a few days, I was promoted to colonel and made chief of the Public Utilities Section, Supreme Headquarters Allied Expeditionary Force (SHAEF), European Theater of Operations. A rank of general was available, but colonel was a field rank, and I wanted to see the fighting and not get stuck in a headquarters position.

Then I flew back to London and was introduced to General Bernard Montgomery's organization of British and Canadian forces. Meanwhile, General Omar Bradley had been transferred from the Italian front to head up the American forces, and I had to familiarize myself with his people and organization. Both men were under Eisenhower. I was able to create a small pool of technical people by transferring some officers in the United States and the Mediterranean Theater and by obtaining some who were already in the United Kingdom training for military government. I also had to learn about the utility conditions in France, and this information was not readily available.

British intelligence flew Louis Alexandre Levy out of central France to work with me. Known as Colonel Leprince in the underground, he was chief engineer of the Department of the Seine in charge of waterways and bridges. He was a bachelor, the grandson of the grand rabbi of Strasburg, and intelligent beyond belief. A lonely man, of small stature, he had great integrity and would go through hell with you. After the war he became inspector general for all of France. Although he had a limited command of English and my French was only barely adequate, we developed a complete understanding. He smuggled out a lot of information including maps of the transmission lines.

June 6, 1944, was D-day, and a small group of us from the Utilities Section followed the invading armies by flying into the first landing strips at Montgomery's forward headquarters. The power systems from Caen to St. Lo had been knocked out by the intense air and naval bombardment that preceded the actual landing. I worked first with General Eustace Tickell, who had been with Montgomery in North Africa and now was in charge of engineering operations for the British and Canadian armies on the north end of the line. In the beginning my activities were centered in Caen. The destruction was terrible. The power plant had been hit, and the tanks were blasting away at each other. One day I had lunch under a railroad car for protection against the shelling, and the British Royal Engineers were under fire as they repaired the plant.

As quickly as supply depots and transportation pools were established, they needed portable generators and distribution lines. So did Montgomery's headquarters and communication centers because more than a million men were ferried ashore the first month. My initial job was to get power moving toward St. Lo, and I got permission to start construction of a line. Tickell protested that he had no steel for poles, and I said we could cut down trees and erect wooden poles. He thought this would not be appropriate for high voltage transmission, but I said we had suitable insulators. When we did reach St. Lo, we found it had been levelled. Nothing more than two feet high was left standing, but the line was built by the engineers and the power did move forward.

The major objective in taking Cherbourg was to insure a deep-water port, and the town was finally occupied on June 26 after severe fighting. Shortly after General Collins accepted the surrender, I inspected the utility equipment and docking facilities. We pulled a destroyer escort with an electric drive into the harbor and used the generators for power. When Cherbourg became fully operational, the allies received about 15,000 tons each day of their initial 20,000-ton requirement. My part was to repair generating and distribution equipment as quickly as possible for the local population of some 40,000 people and to keep the military supplies moving forward.

In July General George Patton broke out of the Cherbourg Peninsula and began the spectacular armored sweep of the American Third Army toward Paris. Although British engineers quickly put a gasoline pipeline under the English Channel, Patton was soon 300 miles farther east. Montgomery was stalled by large German forces in the north which had been deceived by Eisenhower's feint and had remained there still expecting the main thrust of the invasion to come through Calais. He was unable to move forward toward the Channel ports and Antwerp until the end of August.

Eisenhower's plan was to bypass Paris and drive for the Siegfried Line and the Rhine River before winter set in. By not having to transport food, medicines and energy supplies for the civilian population of five million people, he could save enough fuel to permit both the American Third and Twelfth armies to move twenty-five miles a day. The French underground in the city and the Free French division under Bradley's command, however, finally persuaded Eisenhower that he had to occupy the capital. On August 9, General Dietrich von Choltitz had assumed command of Paris with direct orders from Hitler to hold it at all costs, to destroy it if necessary, but under no circumstances to surrender.

The first phase of the scorched earth plan called for the destruction of the water, gas and power facilities. The Germans placed explosives in four of the aqueducts bringing water to the city. And they had two plans for the power plants. They strapped dynamite in the turbines, and the explosions would have delayed full industrial production for two years if the Germans were forced to leave the city. Another plan was to blow up the distribution lines and transformers, which would have cut power for six months if they thought they could return. By mid-August, however, the German supply system had deteriorated so that there was almost no gas and electricity even for cooking, and water was unavailable in some locations. Under these circumstances Eisenhower had to choose between pushing the war immediately into Germany itself, and perhaps ending it by winter, or securing the capital of France and one of the great cities of the world. He selected the civilian over the military requirement.

My big objective was Paris, and Louis Alexandre and I got four cars, loaded them with grenades, rifles, food and several utility specialists. Joseph Shoemaker from The Detroit Edison Company was among them. Despite heavy rain and roads crowded with military vehicles, we got as far as Rambouillet, about forty miles southwest of Paris, before we were stopped at the front of the allied advance.

That night we billeted in the home of the widow of one of the Lazard Frères, and on a battered French typewriter on the second floor, we drew our final plans to get electric power to Paris. I had asked Air Reconnaissance to fly along the transmission lines as they fanned out from the hydro dams in the Massif Central, and the photographs showed where they were broken. French underground fighters had sabotaged them and knew where the insulators had been hidden. During the night, bicycle riders came through German positions from Paris and reported that the power plants were still intact in the city because the underground fighters had prevailed with gold and threats. Louis typed out his orders as Colonel Leprince, and the messengers carried them back through the German lines both to Paris and across the Loire River to Central France to the utility people he knew. He instructed them to bring the copper wire, insulators and transformers from hidden storage areas and to start repairing the breaks in the transmission lines immediately. Once we got them repaired, we also extended the connections to the Italian and Spanish borders.

The next morning was August 24, and we found ourselves sandwiched between columns of French tanks that were banging away at the Germans. I learned that General Bradley had ordered the American Fourth Division and the Second Division of Free French under General Jacques Leclerc to go in by the shortest route through Versailles. Meanwhile, Louis made contact with Leclerc's staff and discovered that the Germans had laid minefields and drawn up tank reserves just ahead. Leclerc was sending a token column to engage them while he moved his main strength to the east so as to enter Paris by the Porte de Vanves and the Porte d'Orléans. We decided to go with this last column because we were pretty sure it would be the first to arrive.

At 2 P.M. on August 25 we started off and reached Paris shortly after the advanced units and were greeted with wild acclaim. People crowded the streets and offered to share the wine and food they had saved for this long awaited day. Leclerc's troops had received the surrender of Choltitz at the Hotel Meurice. When De Gaulle arrived, we went to the Montparnasse railroad station where, as political leader of France, he received the German surrender. The French were overjoyed. Over and over again they said, "Le jour de gloire est arrivé."

We went immediately to 3 Rue Messine and the offices of the Union d' Electricité. Louis knew all the people, and they told us the conditions of the gas, electric and water supplies. We had our rations with us and took supper in the office of Pierre Massey, the engineer in charge. That night I slept on the ground floor of the building with my uniform on and my .45 revolver at my side. There were still strong pockets of German resistance throughout the city and plenty of gunfire.

The next day I met Ernest Mercier, who was the top representative of private industry in the Union d' Electricite, one of the great industrial leaders of France and chairman of the International Chamber of Commerce. We were alone in his office, and I addressed him as Colonel Mercier and said, "I am here representing the Supreme Commander to be of help. In civilian life I am a member of The Detroit Edison Company."

His sternness softened, tears came into his eyes, and he said, "Then you knew Mr. Dow."

He put his arms around me and embraced me. He had been dealing with the Germans for the past four years and did not know what to expect on this day. The people of Paris were in the streets and cafes, however, singing, dancing and drinking.

The next day De Gaulle laid a wreath at the tomb of the unknown soldier at the Arc de Triomphe and then went on foot down the Champs Elysées to Notre Dame Cathedral. There were occasional sniper shots as he marched, but the major German forces had pulled out, and a couple of nights later the Luftwaffe conducted a fire raid on Paris. De Gaulle's principal problem as head of the Free French forces was asserting his authority over the Communist Party leaders, and Eisenhower supported him because he needed political stability behind his lines.

While this intense military and political activity was going on, the physical needs of the Parisians had to be met. I was already hard at work through SHAEF and soon had 2,000 tons of coal coming in daily and stockpiled at the power plants. General P.B. Rogers commanded the area, and I proposed using diesel oil to make cooking gas. He knew that Patton's Third Army needed fuel because the drive to the east had already stalled, but he only asked, "Do you know what you are doing?" I said we did, and he forwarded the request to General Bradley, who approved 300 tons of diesel oil a day to be converted into gas. Just eleven days after the liberation of Paris, we were generating electricity again, and the following day the gas started flowing so that people could cook their food.

For two busy weeks Paris was the focus of my attention, but then the military demands picked up. General Montgomery had finally broken out of his pocket and was moving north along the coast to secure the Channel ports and take the V-1 rocket bases from which the Germans were launching the buzz bombs on London. When he took Antwerp, we finally had a major seaport capable of handling immense quantities of supplies; it also was much closer to the battle lines than Cherbourg. But the harbor and the Schelde River had been heavily mined, and the equipment I needed moved slowly. I was headquartered for a few days at Rheims in the champagne country east of Paris, so I took a jeep and drove north.

Reaching Brussels on September 6, just two days after the liberation of the capital of Belgium, I met Louis de Heem, who was manager of the Société pour la Coordination de la Production et du Transport de l' Energie Electrique (CPTE). He had started to build power stations in the 1920s, and in the late 1930s had begun the national transmission grid. During their occupation the Germans had tried to extend it, but the Belgians sabotaged the lines and not a kilowatt got out of the country. Up until this time the French, Belgians and Dutch had constructed their systems almost independently of one another, and they did not know how to plan together. But they trusted me, and when I chaired the meetings of government and utility men, they laid aside their old national suspicions in a most remarkable way and began to work for their common good. De Heem called his people in, I got some French and Dutch utility men and, of course, we had the experienced British and American personnel who were in SHAEF. The local people knew the war had destroyed so much that none of them had enough to meet his own needs. As soon as we began to lay out a reconstruction plan, however, they saw how they could pool their limited resources and support one another to mutual advantage. This was my first important experience with the international good will and understanding that can be created by the power industry.

These men knew which of their own power plants could be most quickly returned to service and where the supplies were that had been hidden from the Germans. We agreed on objectives and soon had a small scale Controlled Materials Plan going. We cannibalized old equipment, patched the transmission lines, and extended the grid to join the Belgium and Netherland systems. The heavily bombed Schelle Power Plant was a mess, but it had the largest generators in Europe—60,000 kw. When a flash came that the two floating power plants had arrived from New Orleans and were in the estuary of the Thames, I spread a map on the floor to look at our options with General Tickell. I wanted one floater in the Netherlands to maintain the security of the de-watering pumps behind the dikes, and I ordered the *Resistance* to tie into the Schelle Power Plant while de Heem's people proceeded with the restoration of their own machines. I first met Prince Bernhard after he had come over from the Dutch government-in-exile in London and was headquartered at the University of Brussels—before he could go into the Netherlands. He was concerned with many things, and one of them was that large parts of the country could be flooded. Through SHAEF we were able to bring in officers experienced with pumps and de-watering operations.

I ordered the *Seapower* to proceed up the Schelde River regardless of the ratchet mines although I sweated out the move for thirty-six hours until it tied up at Langerbrugge near Ghent. The two floaters together added 15 percent capacity to the Belgium system and helped the operation of the port facilities of Antwerp to receive our steadily increasing flow of supplies. I wanted to share a small part of all this with my family, so I added a discreet sentence to an airsheet to Jane, who was now ten years old. "I have seen today what you christened in Pittsburgh two years ago."

By mid-September the allies had control of the Limbourg coal mining regions of Belgium and the Netherlands, but just to the east the battle of Arnheim had begun to produce terrible casualties. These were the largest mines in Europe, and although the Germans damaged the electrical systems they did not destroy them because, as they told the Dutch, they expected to be back within three months. Some 25 percent of the Belgium capacity and 75 percent of the Dutch capacity were damaged or destroyed. I was in Antwerp studying an inventory of the power system when the Germans ordered the operators out of the Merxem Power Plant and shot them as they came through the door. Tickell turned to our group for advice on the coal mines, and I said to him, "I will get the power, General, if you can build the transmission from Merxem to Roosendaal to operate the pumps."

He said, "I'll build your goddam line for you," and he did.

While the artillery was still firing, we got the pumps started again and kept the 2,500-foot mines from flooding. With these local supplies to augment what was coming from England, the factories began to start up just behind the lines, and some of the first steel produced was used to construct the bridges for the final crossing of the Rhine.

Even as the War Production Board had to be organized in the United States, so production had to be started again in Europe. SHAEF was organized by missions—in Belgium, France, Italy and the Netherlands, and I had officers in each. After the initial Normandy landing my principal task as chief of the Public Utilities Section was to get the reconstruction started and to keep it moving forward. This required bringing knowledgeable

Europeans together, agreeing on goals and cooperating with one another. The U.S. Army Corps of Engineers and the British Royal Engineers did most of the work to meet the immediate military needs. But the local people did the permanent rebuilding and they operated the systems. They knew what needed doing and how to do it. Some of them began to refer to all this as the Cisler Circus because we had so many activities going at one time. But my role was one of coordination since I had key information on shortages, knew where supplies might be located, and had the overall conception of total requirements.

On one occasion I called in a group for consultation and the allied representatives were surprised when I included a couple of Germans. They were still our enemy, and one of them got sufficiently confused to give the Hitler salute. He was very embarrassed about it. But I felt they had to come in as early as possible for the good of European recovery. And the national groups knew that we had to move power from wherever we could generate it across whatever transmission lines were available.

I always brought people together without using military authority or commanding them. We would first reach agreement on what was a good thing to do, and when this occurred people worked together willingly. When I reached Germany, I shared my food with the local utility people. Some said this was a court-martial offense, but many of these men were close to starvation. Their plants had been blasted to pieces, and they recognized they were a beaten nation. I was there to help re-establish the European systems, and this could not be done without local people who had a sense of their own worth.

The allies first breached the Siegfried Line at several points, but the crossing of the Rhine River was delayed for six weeks until supplies could be brought through Antwerp. We had seven armies on the Continent and the logistic demands were enormous. To help meet them the industrial capacity of western Europe had to be restored as quickly as possible, and this required the coal supplies of the Saar. In World War I the Germans had flooded the mines, and it had taken two years to get them back into full production.

General Patton had crossed the Moselle River near Metz, and there was intense fighting along the whole front that had settled into a slugging match. Then the Battle of the Bulge erupted, and his tanks were occupied with containing the German offensive. I felt, however, that we had to get into the Saar region, and we used our friends in the underground to help. So even though Louis Alexandre had a terrible cold and was running a fever, we moved east from Metz. At 10 P.M. I had to walk in front of our cars with a flashlight because of the dense fog. The entire line of battle was ablaze and it was very confusing. But we got hold of the power plant at Carling and another nearby before they could be destroyed. As a result of this and other advance activities only part of one mine was lost.

When the drive across the Rhine brought the armies as far as Frankfurt, it was important to stop any flow of electricity into Germany, and Lauffenburg was the last power plant on the river that was serving them. It was located right on the border, so that half of it was in neutral Switzerland. General Clay asked me to put on civilian clothes and negotiate with the Swiss. This did not seem right to me, however, so I asked to proceed in uniform. I went in at night through our lines and entered Switzerland by walking across the turbine room. The Swiss had a financial agreement to supply power to Germany, but they were willing to let us switch it over to the transmission lines through Colmar and the Belfort Gap and into Paris.

Allied Military Government

Once in Germany, we found the power plants had been bombed by air attack, but the total system had been the most advanced in Europe and was the only one comparable in sophistication to the United States. I followed the armies, reorganizing the local utilities as we proceeded, and finally went into Berlin with General Clay. The city had been badly smashed. I visited the bunker where Adolf Hitler and Eva Braun had died. Everywhere there were dead and decaying bodies. Once you get this smell in your nostrils you never forget it.

When the fighting was over, there was much to be done, and we knew the power system had to work as a unit regardless of the American, British, French and Russian zones. At first our contact with the Russian technical experts gave us high hopes, but soon the political members appeared and took control. Then there would be long hours of discussion, and agreements reached one day would take on a different meaning the next. But I had been working with representatives of the U.S.S.R. since the early days of lend-lease, and they knew I had some understanding of their fears and of the immense destruction in their country. Despite the many political differences, the electric power system was interconnected, and it began to operate as a single unit.

During these months I had gradually evolved a plan for the revitalization of Europe. Henry Morgenthau, who was secretary of the treasury and a powerful man in the Roosevelt administration, wanted to turn Germany into an agricultural state, but I believed there would be no enduring peace unless the nations were economically strong and had pride in their accomplishments. I had learned that there was great strength in government if people knew how to use it. We quickly got the mines operating again, cement and steel and other industrial products coming out of factories, and we had started to rebuild Europe. By early 1945, the consumption of electricity in Paris was 38 percent higher, and throughout France was 10 percent higher than in 1938, the last normal prewar year. Belgium suffered greater damage, but generation returned to 80 percent of normal. The Netherlands sustained the greatest destruction, but it also reached 50 percent of normal output by March, 1945.

In Europe I learned again what I had discovered in the War Production Board. When a nation can agree on goals, it is capable of great achievement. During the war our adversaries were the axis powers, and since our attention was directed toward defeating them we did not spend all our efforts confronting one another. American society has made significant advances through competition and opposition of organizations, but I became more aware than before how much can be done through cooperation. We stress this within organizations. Athletes emphasize teamwork even while they train to oppose other teams. It is an effective balance between competition and cooperation that is hard to find, and we reached it with ourselves and with our allies as never before. The value of cooperation appealed to me. It suited my character.

On July 15, 1945, I was on the air field in Frankfort when Eisenhower returned to the United States to become chief of staff. SHAEF was closed, and the Office of Military Government, United States (OMGUS) took over with General Lucius Clay in charge. In September, Louis de Heem came to Berlin to ask for additional help in getting his pre-war work with the Belgium grid started up again. Ernest Mercier, who headed CIGRE, the international high tension group, also wanted to participate. Then Harold Hartley came from London to ask if I would help the World Power Conference pick up its activities. As a result, I organized the first postwar energy conference in western Europe. Invitations went out to the principal power representatives to meet in my billet. I scrounged around for food and got principally Spam and whiskey which introduced most of them to bourbon and rye. The Russians sang their songs. At midnight the young girl who was the translator went home, but we still had plenty of liquor and Louis de Heem had a notebook with enough Russian phrases for us to get along. The party broke up at around 2:30 A.M. and several groups had to navigate by the north star in order to find their quarters.

On September 17, the day Winston Churchill was voted out of office, we had our conference. This meeting accelerated the rebuilding of the existing organizations and led to the formation of the Public Utilities Panel of the Emergency Economic Commission of Europe to coordinate the electric power facilities on an international basis.

In November, 1945, I finally returned to Detroit and the duties I had assumed two years earlier. Whereas before the war, I had accepted utility regulations as part of the public interest, I had now learned to work with government during the period of the most absolute control over personal and business life in the whole span of American experience. I had seen what happens to advanced industrial nations when their utility systems break down. The discovery of new processes such as artificial rubber production, and others in the chemical sciences and engineering, brought whole new industries into existence and changed the allocation of scarce resources into the management of abundant supplies. Nuclear fission was one of these new processes.

The war in the Pacific ended with the explosion of atomic bombs over Hiroshima and Nagasaki. With this demonstration the world knew that there was another source of primary energy and a wholly new process of converting that energy to the peaceful uses of mankind. Profit continued to be the measure of successful business enterprise, and we learned that in addition to victory we had also achieved economic growth. What remained to be resolved were the political and social issues in maintaining this growth and in extending it to all the peoples of the world.

CHAPTER III

BOYHOOD AND PREPARATION

I was born near the Cisler farm October 8, 1897 in Marietta, Ohio, at the confluence of the Muskingum and Ohio Rivers. The town had been established by the Ohio Company right after the Revolutionary War by old soldiers who knew how to pick a good trading site. Although the surrounding country was largely given over to farming, Marietta was served by two railroads and the rivers. It was also a pioneer area in the developing oil industry. Marietta College provided intellectual and cultural stimulation and embodied the idea of higher education by its very existence. The townspeople and near neighbors could enjoy a variety of music, art, athletic and lecture programs as spectators. They could also consider sending their children to college.

My family's belonging to the middle class gave me additional advantages. My mother came from a family of small businessmen who were touched with the spirit of adventure. Her father had emigrated from Yorkshire, England, where he was in the wool trade, and he followed this business in Pennsylvania. My maternal grandmother died when my mother was only four years old.

Of her four sons, the oldest, William E. Walker. went first to Minneapolis to set up a business and then moved farther west to Great Falls, Montana. He dealt in hides and tallow, and a large part of his business was buying wool for a Boston firm. In time he brought his three brothers to join him in Montana, and in the beginning they lived in a one-room shack, did their own cooking and ran the business from a corner of the freight station of the Great Northern Railroad.

My father, a graduate of the University of Pennsylvania, was a medical doctor. I had an older and a younger sister, but when I was four years old my father and mother were divorced. He remained in Marietta in the general practice of medicine and later remarried. My older sister, Frances, remained at the Cisler farm with our grandmother. My sister Anna and I returned with my mother to her sister and family in Wallingford, Pennsylvania, where for several years we lived in a series of rented homes. Finally, my Uncle Will bought a thirty-three acre farm in nearby Gradyville for us because his father had said on his deathbed, "Willie, always watch over Trot." My mother, Sara, was the youngest of six children, and she never did anything unless on the trot. There was a good deal of bitterness between my parents, and I was separated completely from my father. So even though I came from middle-class stock in middle America, I also had some liabilities that shaped my life.

Looking back over seventy-eight years, I realize that growing up is a matter of learning how to manage imbalances in one's own development. I became an engineer and a businessman, and I attempted to create an order for myself in terms that I gradually grew familiar with and could make productive. Just like nations, I had four elements that needed to be brought forward in a balanced way. Around me I had people, resources, technology and energy.

In addition to the people in my own family there were neighbors. Gradyville itself was no more than a crossroad about two miles from the nearest Pennsylvania railroad station, which itself was about twenty-five miles southwest of Philadelphia. There was a small hotel, a creamery, a general store with the post office and a one-room country school. The surrounding population was made up mostly of prosperous Quaker farmers and a few bankers, lawyers and railroad officials who worked in Philadelphia. They also owned some of the farms which were run by hired hands—chiefly Italian and British immigrant families who were provided a house, food and a job. When they had earned enough money, they bought farms and dairy cattle of their own.

In the early mornings the farmers took their meat, sausage, scrapple and vegetables by wagon to the railroad station and sent it to the Philadelphia market packed in ice. The milk was usually sent along in ten-gallon cans, but some of it was delivered to the Gradyville Creamery to be churned into butter. I can still recall the sweet clean smell of the dairy, and I learned to enjoy the sour taste of buttermilk, especially with the small chunks of butter floating in it.

40

Later in the morning, the businessmen commuted on the railroad to their jobs. There was also an interurban trolley nearby—a single car holding about thirty people, with an energized wire overhead, a hand-regulated transformer to control the power, and hand brakes. The locality had a cultured, intellectual environment. Swarthmore College was only seven miles away, and Haverford College was nearby. Many of the people were Quakers, and almost everyone I knew was a member of some organized religious group. Although I, myself, did not attend church regularly, I was brought up to respect the beliefs of others and to appreciate the values of honesty, thrift, hard work, concern for others and personal integrity that were preached from the pulpits on Sunday and practiced everyday.

In general, I was encouraged and praised rather than scolded and punished. There must have been serious differences among the adults of Gradyville, since many were people of strong convictions, but what remains in my memory of those first years are the people I could trust. The men would speak to me on their way to the station as did the farmers as they took care of the animals. When I was thirteen, I began to deliver mail between the store and station each day. Everyone knew one another by name, and they knew me.

One neighbor was especially important in my youth. John Murray was a close friend of my Uncle Will, and the Murray and Walker families lived across the street from one another in nearby Wallingford. Uncle John's father had been the gardener for Dr. Horace H. Furness, the foremost American Shakespearean scholar of his day. After the father's death, the family continued to live on the property, and Dr. Furness sent Uncle John through Swarthmore College to study civil engineering—the best money he ever invested, he said. John Murray was a tall, remarkably fine looking man, and he was always very kind to me. He remained a bachelor all his life, and I called him Uncle John and knew his sisters as Aunt Mary, Aunt Nan, Aunt Julia, and his mother I called Grandmother Murray. There were other families close to us during my formative years, but I especially admired him for the way he looked after his family. He was associated with the Pennsylvania Railroad, and this was the great period of steam locomotives. He talked often about the building of the Hudson River tunnel, the Hell Gate bridge and the Pennsylvania station. I am sure he influenced my decision to become an engineer because I thought anything he did would be important for me to do.

The resources of my early years were neither abundant nor stable. In fact, unless everyone in our house worked hard and lived prudently, life could be quite precarious. When I was seven years old, Uncle Will returned from Montana for a few months and had a new house built on the massive stone foundations of an older one that had burned about twenty-five years earlier and which, with various changes, had existed since about 1750. There were four bedrooms upstairs, a kitchen, dining and living room downstairs and a basement. I was responsible for the coal furnace and the ashes. There were two fireplaces, oak floors and fine chestnut-beamed ceilings. The kitchen stove burned wood, and I split the kindling. Ice was cut from the local pond during the winter, and we packed the blocks in straw in our ice-house. We had a garden that I helped cultivate, and Mother put up jars of vegetables and fruit for our table. I fed the horses, milked the cows, sheared the sheep and collected the eggs from 500 chickens. I got buttermilk from the dairy and mixed it with grain middlings, and the pigs would jostle for position at the trough, squealing and snorting as they guzzled.

A wooden windmill pumped water from the well to a tank in the loft of the barn for our running water, and this was the first equipment that made me aware of mechanics. My earliest contact with electricity came from an alarm system in the barn. It was run by battery, and I bought ten cents worth of salimoniac powder to power it, rigged it to a doorbell and then ran a line to the house. We had no electric lights until later, but there was a gas lighting system in the barn. We used kerosene lamps in the house. I got my hands covered with soot many times as I cleaned the chimneys, and this was particularly distasteful in the winter. There was more soot then because the nights were longer, and it was hard to get the carbon off because the water was cold.

Mother had been taught to be a seamstress, and she could make her sewing machine hum by pumping the foot pedal. I had no mechanical toys, but one time I disconnected the drive-belt and attached it to a row of empty thread spools that were on nails. The treadle made them spin at a great speed. Later I bought a motor to run Mother's old sewing machine, and as soon as I could afford it I bought an electric model for her.

No one that I knew was really hungry or cold, but occasionally there were serious difficulties to contend with—sickness, poor crops and bad weather. Mother was persuaded by her brother Herbert to invest her small inheritance in a big flock of sheep in Iowa, and they were largely wiped out during an unusually severe winter. So her little financial resource was gone, and I began to feel an even greater sense of responsibility for her. The home had become very important to me. I still own it, and the oak board still hangs over the hall fireplace with the hand-lettered message:

Each man's chimney is his golden mile stone, the
central point from which he measures every distance.

When I was thirteen, my mother married Henry Carter, who had been born and educated on the Isle of Wight off the south coast of England. The Carter family was in the paper-making business, and this brought some financial stability into our lives. I had had to grow up fast, and by this time I felt quite adult. But because of my responsibilities, I had been absent from school a good deal and had fallen behind. I really did not begin to attend the one-room schoolhouse regularly until I was nine years old. Now I could do so and I finished the elementary grades. In time my three new sisters, Jane, Hester and Sara, whom I cherished, also started their formal education in the same nearby school.

Education has been in my life what technology is to the life of a nation. The first day my mother took me to school to meet the teacher she said, "Whatever may be your good fortune to do in life will largely depend on the success with which you acquire an education." I have tried to keep learning new things and to remain current on those areas I needed for my responsibilities as a businessman. From a purely practical point of view an education is one of the most important tools a man can put his hands to. Since it was also a part of my youthful heritage, it is something I treasure for its own sake.

I found satisfaction just in knowing more and more about both the social and physical sides of the life I was experiencing. Over the years, I came to believe that education is one of the primary needs of the world. With formal training we learn to cultivate our minds, and it is through the mind that we manage our own lives and the planet Earth. In 1975, there are about four billion minds throughout the world that will determine whether mankind makes progress toward peace, economic abundance and intellectual inventiveness, or retrogresses as happened in the Dark Ages.

The one-room schoolhouse in Gradyville was a strong influence in pointing me in the direction I have travelled. All the elementary school-age groups were there, and the older children helped the younger. They were of English, Irish and Italian descent; they were black and white; they were from new immigrant and old settler families.

Many boys went to work after finishing elementary school, but fortunately it was possible for me to continue my formal education. Delaware County held examinations to qualify students for going to high school and would pay the tuition at the school of their selection. I chose to go to Westchester, nine miles away. I could combine schooling with carrying the mail in the wagon and so earn some additional money. Bicycle clubs were popular, and the members thought nothing of riding to Atlantic City a hundred miles away. This got me into a form of exercise that I continue to this day. Except in the worst weather I still get up winter and summer at five o'clock and ride five miles before breakfast. Whenever I travel, I try to take a bike with me or borrow one at my destination. I must have ridden 100,000 miles, and it seems a natural thing to do because I started so early.

In high school I enrolled in the commercial section since there was some thought I would join my uncle in the west. At first I took courses in penmanship, typing and commercial geography. At the beginning of the second year, however, Guy Chipman, the principal, and a teacher named Reid Henderson, who had graduated from Penn State University, said they had been watching me and thought I ought to take the college prep course. "We think this is where you'll do best," they said, "and if you're willing the teachers have agreed they will make up for your first year by giving you instruction after school hours." In this way I got four years of Latin and English and had excellent courses in chemistry and mathematics.

These men and women who were my teachers reached into my life and gave me new opportunities. By the time I graduated I had twenty college preparatory courses where only fifteen were needed. And because this high school had such a good reputation, Cornell accepted me into the engineering school without a question.

44

I have always tried to put back into the educational system some of the good I received from it. For a number of years I was an active member of Cornell's board of trustees, and I have been on a number of other boards including that of Marietta College. In the same way, I have tried always to include leading educators on boards and commissions in which I have had a role. When I became president of The Detroit Edison Company, I urged that Dr. Harlan Hatcher, then president of The University of Michigan, be named to the board of directors, and he was. Dr. Robert Bacher, an eminent nuclear scientist during the war years and later provost of The California Institute of Technology, also became a member of the Edison board. In most major episodes described in these reminiscences, faculty members have taken a significant part. Some of them have possessed the most creative minds of this century.

I have also tried to respond to requests from teachers in the elementary and secondary schools, and I have been active in both national and state programs to encourage private support for them. Few other nations have ever set themselves the goal of making knowledge available to all their citizens and of taking them out of the work force for sixteen or more years until they reached the limit of their ability or their desire to continue. Indeed, the goal turned out to be possible only because machines had freed a large part of an individual's life from the hard toil of subsistence living.

If family, community and education were important elements in my growing up, so too was youthful energy. I was blessed with a strong constitution. There was nothing I would not do to relieve my mother from her burdens. I prided myself on being able to do a man's work, and this meant physical labor in the fields and caring for the animals. In high school I played on the football team, and I was on the track team at Cornell. Although I had the usual boyhood diseases like measles and chickenpox, they were not serious, and today I seldom suffer even as much as a cold. I am still active over an eighteen-hour day and a seven-day week, because I find the rhythm of wakefulness to be exhilarating. I am one of those fortunate people who can manage on five or six hours of sleep, and this has added greatly to the time available for more interesting matters. The amount of personal energy surely helped balance the other elements that went into the design of Walker Cisler's life.

But there were also other forms of energy available. I learned

45

from practical experience with our hand-fired furnace that anthracite coal was a denser form of energy than wood and yielded more British Thermal Units (BTU) to heat the house. The railroad used bituminous coal. I could not imagine fuel tenders large enough to hold wood for the locomotives of those days, although pictures showed smaller wood burners thirty years earlier. In my senior year in high school the family bought a car, and I used gasoline as a still more concentrated form of energy. As coal had to be mined, so oil had to be upgraded in a refinery. But electricity was the most sophisticated of all forms of energy and the most flexible. For a few minutes the average man can turn a generator by arm or leg muscle enough to light a thirty-five watt bulb. In a year he can do work equivalent to about seventy kilowatt hours. A skilled industrial worker uses about 14,000 kwh annually, so he is as productive as 200 workers. I accepted such benefits very quickly. All of these forms of energy brought me into the way of life of the urban industrial world in which the advanced nations find themselves today.

Cornell University

By the time I graduated from high school in 1917, I knew I wanted to be an engineer. Cornell had such a great reputation that I wanted to study there, and to help make it possible, my mother sold twenty oak trees for $250 and I got a campus job. Since I was already two years behind my peers, I settled down to get the best education I could. I also joined the Student Army Training Corps because the military wanted the top third of the engineering class deferred from active duty to finish their technical education. But at the end of my freshman year I enlisted with a group of students who were determined to see action.

The following fall was terribly wet. Rain poured down and Camp Zachary Taylor, where I was being trained, was flooded. It was also wracked with an epidemic of influenza, and many of the boys burned up with fever and died. The virus did not affect me, and I was commissioned a second lieutenant. When the examining board learned I could handle horses, they put me in the field artillery. Although I gained my first management experience in administrative duties with the army, Armistice Day, November 11, 1918, came before I could get to Europe.

The following February, I was mustered out of the army too late in the academic year to return to Cornell. So I got a job with the Pennsylvania road department as a quality control inspector and came to know about asphalt, concrete, reinforcing rods and soils. With the wages added to my army pay, I went back to my studies in the fall. I specialized in mechanical engineering, but I also took the electric power option in the senior year. And I was able to return to the road department each summer for employment and further practical experience.

There were other activities besides study. I organized the Track Club. Later, I became a member of the honor societies Tau Beta Pi and Phi Kappa Phi, and I also joined the student chapter of the American Society of Mechanical Engineers and became its president.

During the War, membership in the Society had dropped, so a team of us decided to host a cider and doughnut party to sign up new members. My roommate, Frederick W. Utz, whom we called Fritz, borrowed a Model T Ford, and we drove over West Hill to a cider mill to buy a large keg. On our return we bought several dozen doughnuts, but when we got back to University Hill the motor died. I cranked until my arm ached. It would sputter a couple of times and die out again. Fritz cranked. Nothing that a couple of mechanical engineering students tried was of any help. Finally, Fritz took the doughnuts, I heaved the keg up on my shoulder and we rode up the hill on the open back platform of the trolley. We trudged the remaining distance to the picnic and were hot and thirsty when we reached the group. That was the most refreshing cider I have ever tasted.

Many important people in the engineering profession gave lectures at conferences in Ithaca. Alex Dow, president of The Detroit Edison Company was not a college graduate, but he knew many of the faculty and he visited the university several times. A number of the principal managers of Detroit Edison were Cornell graduates. Vice President James W. Parker was one, and I heard him speak. Paul Thompson was another; he was chief engineer of power plants in Detroit. C. F. Hirshfeld had been on the faculty, and Alex Dow had persuaded him to go to Detroit to establish the engineering research department. I used his textbook in my courses. These men and others from Detroit were active in the engineering societies, in research and in publication.

In my senior year I was interviewed by a number of companies, but I was really interested in the electric utilities. They had good beginning salaries, and Public Service of New

Jersey paid the top dollar. A year earlier it had started a cadet training course that was new in the industry, and this promised to broaden the experience of a beginner very quickly. As a result, six of the men in my class decided to go to Newark, and I, too, chose to join the New Jersey company. This employment had the extra advantage of being reasonably close to my mother and sisters.

I realized later that the first twenty-five years of my life were lived during a critical period of history. There were few labor-saving devices in the opening decade of the century. Women worked around the house and garden much as they had always done for centuries, preparing food, making clothes, keeping the family organized. Hard physical labor was required in shops and mines from morning 'til night. Men worked in the fields the same way. Scientific agriculture had been introduced earlier and had begun to improve seeds, fertilizers and livestock, but the great advances in farming technology still lay in the future when machines would largely replace the muscle of men and animals.

In the second decade the United States emerged as a world power and as an equal to Europe in industrialization. Electricity could be applied to all manner of machines and appliances, and it could do unlimited amounts of work. It was the best servant that had ever been perfected. As the benefits spread, more people developed an entirely new way of living, and the United States became the most productive country in the world.

The Young Engineer

I liked the philosophy of public service. As a businessman I have never been strongly attracted to earning a large salary, and I know a surprising number of others who have not been interested in amassing fortunes. The challenge of innovation and accomplishment has been more rewarding. I was competitive in engineering fields, in trying to design better machines and energy systems, and I understood that it was necessary for the company to have money in order to build them. The year I graduated I was twenty-four years old, the economic expansion of the 1920s had begun, and the use of electricity had really taken off after World War I. Even as I knew what I enjoyed doing, it seemed to me that the power industry was where I wanted to be. Here I felt I could turn my talents to making life easier for people.

Public Service of New Jersey was a conglomerate made up of an electric company, a gas company, and a railway company which operated ferries and an interurban line between Newark and Trenton. There was also a Public Service Production Company, which was an engineering and construction operation that built power plants and underground cable installations and even placed a cable under the Mississippi River. In addition, it interconnected with the Philadelphia Electric Company and was involved with United Gas Improvement of Philadelphia. Such a wide variety of activities brought me into contact with many technical problems.

During my two-year cadet period it seemed important to augment my engineering with some business and economic studies. New York University offered courses in these subjects, and I went to the Fourteenth Street campus three nights a week. These were lecture-style courses, however, and I wanted more discussion. So a group of us in the company got together at night in the Public Service headquarters to take the Alexander Hamilton Institute extension course. Each of the twenty-four volumes concentrated on an aspect of business such as accounting and marketing. The authors were people of distinction, and we took up a book a month for two years. One of our group was assigned to act as the discussion leader of each volume, and since we were with our own people we did not want to slight our studies. Some of the non-technical people in the executive training program who had graduated from Princeton and Harvard also joined, giving us the advantage of a very broad range of views. Economic thought was relatively simple in those days before John Maynard Keynes and others began to challenge businessmen with new perceptions of how the free enterprise system worked and could be controlled. Still, I had to go far beyond the engineering economics I had studied at Cornell, and I met people with different ways of looking at the world.

At the end of my cadet period I chose to concentrate on the electric side of the business and became a test engineer with the Kearny power plant that was then being built. I visited the Trenton Channel power plant of The Detroit Edison Company because it was quite similar in design but was farther along in construction. Here I renewed my acquaintance with Professor Hirshfeld and learned of the research he was undertaking. When Kearny came on the line, I was sent to an older plant in Patterson as chief engineer, and while there I was given permission to set up a generation research committee with the charge of doing some advanced technological planning. The committee was composed of a small group of young engineers, and we all had our regular jobs to do. But we operated as a kind of think tank—very elementary by the standards of the 1970s—because we did not have computers or advanced mathematics—and we were given a good deal of freedom to pick our own subjects. In the plants, for example, there were voluminous reports that had been compiled ever since the early days when generators were powered by reciprocating engines. There were data on bearing temperatures, steam pressures, tons of coal burned. We organized this mass of information and compressed it into a useful form. We organized preventive maintenance schedules on the basis of experience with the machines rather than simply because the superintendent thought it was about the right time.

These veteran engineers were practical men who had lived with their machines and knew a great deal about them. What our group did was to systematize statistical data and link it with the judgment of operators. We studied the economic as well as the technical advantages of what we wanted built into the plants and were able to introduce new equipment where operating efficiencies justified the increased expenditures. In brief, we were actively involved in accomplishing what most young engineers hope do to in their professional lives.

Although I spent much of my time during the 1920s learning the technical elements of my work, I was also exposed to the social changes being produced by industrial expansion. More and more people moved into cities and suburbs so that the small rural community where I had grown up became part of a metropolis. One of the popular songs following the first World War asked the question, "How you going to keep 'em down on the farm, after they've seen Paree?" Women got the right to vote and were freed in many ways from the roles they had traditionally played. In addition to teaching school, they were becoming secretaries and bookkeepers, and they began to enter the work force of the business world. The motion picture industry boomed, and movie palaces were built for several thousand people at a showing. Radio became a popular form of entertainment in the home, and book and magazine publications flourished. Information of all kinds began to compete for attention as the new advertising industry grew. There were flappers and prohibition gangsters who made headlines. Charles A. Lindbergh became the heroic Lone Eagle because of his solo flight across the Atlantic. Although dirigibles were already crossing the ocean and exploring the feasibility of mass air travel, Lindbergh's exploit emphasized the development of the new form of transportation by air. There was new literature, art and theater—a renaissance of the creative spirit. Scientists became popular and widely read. The bases of contemporary biology, chemistry and physics were being expanded, and thousands of students were entering these fields as the university populations grew larger and industry began to develop commercial applications of science.

Government regulation of the power industry first began in the opening years of the century through company negotiations with local town and city councils. Electricity, along with gas and water, required the use of public streets for the economical construction of distribution lines. Such services were recognized as being so vital to individuals and the community that the public interest was held to be higher than in most other business enterprises. In some municipalities the local government provided the services, but in many others, private corporations negotiated franchises to do business subject to restrictions.

The growth of the built-up areas finally outran the ability of industrial towns to annex their suburbs, while new population centers grew jealous of their autonomy and incorporated in rings around the core cities. The growth of power companies was even more rapid as customer demand for electricity tripled between 1913 and 1919, the years surrounding World War I. It doubled again in the next seven years, and some power companies began serving thousands of square miles connecting several large cities and many towns. When the movement toward metropolitan and regional government did not develop beyond the large-city concept, regulatory control gradually passed over to state commissions. As early as 1907 Wisconsin and New York recognized that electric lines had begun to run beyond municipal limits, and gradually all states began to accept responsibility through railroad or utility commissions to establish uniform accounting procedures.

Public Service was the largest utility in New Jersey, and it operated both statewide and outside state boundaries. As prices dropped and service improved, the regulatory agencies focused largely on making sure that good business practices were observed.

51

Through all these events, and over a period of ten years, I grew from a financially insecure country boy who had to struggle to balance his small world into a professional engineer and businessman living in a quite different environment.

The Depression

In 1929 the nation celebrated the fiftieth anniversary of Edison's invention of the electric light, and the power companies had special reason to celebrate. Herbert Hoover, who was known around the world as an engineer, businessman and humanitarian, was President. It seemed as if nothing could stop the forward momentum of the industrial revolution. Then came the great depression of the 1930s.

The depression years affected everyone. The art and literature and theater of the day featured social realism, and everywhere one saw a picture of gloom. Industrial people found themselves without jobs and hourly-rated workers turned to unions for protection. Farmers could not sell their products at a price that covered their costs. Their plight was worsened by a series of severe droughts. One of the most disturbing events of the period was the migration of thousands of people who left their farms in the dust bowl and just moved on. The country was unprepared for dislocation on this scale. Government services at all levels were inadequate to meet the suffering of people who felt the economic system was falling apart.

This movement of people threatened the food supply of the nation, and President Franklin Roosevelt took many actions in the 1930s to slow it down. The Rural Electrification Authority (REA), the Tennessee Valley Authority (TVA), and the Bonneville Authority were established by acts of Congress to help farmers get the advantages of electricity that they needed for efficient operation. These actions also appeared to some people to be politically motivated, and they certainly did put the federal government into direct competition with investor-owned industries. Henry Wallace, as secretary of agriculture, appeared to represent some kind of governmental interference with independent farmers. Secretary of the Interior Harold Ickes seemed to many businessmen to represent government interference with free enterprise.

The business failure of the Insull group of utilities of Chicago was the most highly publicized collapse in the power industry. The economic practices of a few companies, along with the depressed capital market, caused the Congress to pass the Securities Act of 1933, establishing the Securities and Exchange Commission (SEC) to protect investors and help American industry raise money for plant expansion through the sale of stocks and bonds. When the Utility Holding Company Act was passed in 1935, the SEC was authorized to administer it.

The Utility Act was directed toward dissolving the holding companies and controlling the kind of financial manipulation that they practiced. Autonomous organizations like The Detroit Edison Company were not affected because they were geographically integrated and were genuinely owned by the stockholders. Public Service of New Jersey, although it was made up of a number of companies, had not watered its stock, so it was also exempted from the Act. Although my associates and I were not affected directly, we were exposed to all the publicity surrounding these events. Many of them were closer to the financial, marketing and regulatory activities than I, and less involved in the design, engineering and construction of power plants and transmission lines. They felt threatened by the political forces that wanted to nationalize the industry. There was a good deal of hostility. I understood this concern, although such political issues were not part of my daily work.

President Franklin Roosevelt moved the political question of regulation from the state to the national level and reopened the public vs. private power controversy. In practice he did not have a sustaining conviction about nationalizing anything, although there was plenty of strong rhetoric flowing around him. Events showed that he was willing to use the REA and TVA to help a limited number of people and localities through hard times. But electricity alone was not going to keep young men on the land. As small farms became unprofitable, it was necessary to cultivate more acres by machinery to support a family. Perhaps electrification helped slow down the migration, and certainly it improved the productivity and amenities of the individual farmer, but no national competition from government-owned power emerged. I knew that all the federal resources poured into TVA and Bonneville produced good systems, but they were no better than Public Service of New Jersey had built at no cost to the taxpayer. And our generating equipment was superior to most of the REA machines. I had become an experienced engineer and I understood the comparative value of the products.

Public Service of New Jersey was not directly engaged in the public-private power conflict. Franklin Roosevelt called our Chairman Thomas McCarter to Washington to advise him on what to do. As I saw it, Roosevelt was more concerned with stimulating the economy than with regulating it. The Holding Company Act was necessary to protect the industry from excessive financial speculation. The federal power projects were demonstrations of what the government might do rather than would do. As a result, although I did not like interference with the obligation of management to make decisions, I did not feel hostile toward either government engineers or the politicians. And when the demand for electricity began to increase after 1935, the investor-owned companies picked up their rate of growth, and the percentage of government ownership in the industry dropped steadily.

President Roosevelt's reassuring statement that we had nothing to fear but fear itself was substantially true. But fear, whether from real or imagined causes, is one of the most powerful motivators of human action. People seldom act their best under these circumstances and frequently act their worst. So I learned that there are imbalances in national emotions that also require careful management. Everywhere I went I saw the results of idle machines. Men and women were forced back into hard physical toil, marginal subsistence and a precarious future of the kind I was familiar with in my youth.

The impact of the depression was felt very quickly in the power industry. As factories shut down, industrial customers used less electricity. As fewer goods were bought and sold, commercial customers cut back on the size of their stores or the number of hours worked. Many of them went out of business. As wage earners lost their jobs or found their pay cut back, they used less electricity in their homes. All this happened very quickly, and the graphs that charted the usage reflected precisely what was happening. In 1930 the exponential growth rate suddenly stopped, and during the next three years the use of electricity actually declined. Not until 1934 did the curve start up again, and it was not until 1935 that the level of consumption equaled that of 1929.

This six-year dip was a sensitive barometer of the welfare of the nation. The whole economic system had become sufficiently dependent upon electric energy that every change was reflected in the patterns of use.

Like everyone else I found my pay reduced, but at least I had a job. Although the construction of power plants was slowed down, our economic studies showed that it was worthwhile to retire older machines and replace them with more efficient new ones. I was in Jersey City as chief engineer of the Marion Generating Station, but I also continued my special assignment in the general offices for advanced planning. We were looking at even newer ways of producing energy.

Our generation research committee made studies for pumped storage in which water would be raised to a reservoir high in the hills at night when there was plenty of power to run the electric pumps. It could then be allowed to run down in the day when the demand was greater and to turn the generators to produce the electricity needed in homes and factories. The technical calculations of the costs and savings were made in my office. Although we did not build an installation because we had excess productive capacity at the time, the studies proved that this would be an economical form of generation.

In addition to water we also looked at wind power to produce electricity. The conventional windmills were far too inefficient, but Julius Madaras had taken out patents on rotating vertical cylinders. At Detroit Edison, Hirshfeld developed the idea of hooking forty flat cars together and running them around a circular track. Mounted on each car would be a rotating cylinder ninety feet high and twenty-two feet in diameter. The windpower from the rotating cylinders would push the car and drive a generator capable of producing 1000 kw. The track would be a mile in diameter, and each cylinder would have to slow down at the half-way point, then stop and reverse the direction of its spin. The rotating cylinder acted like an air foil. The wind pressed on one side to turn it, but it created a vacuum on the other side that greatly increased the force, and this was the Magnus effect. The resulting output was ten times as great as that created by a conventional sail of a boat or a windmill.

Hirshfeld organized a seven-company group to put in research and development money, and we built a full-scale cylinder at the Burlington station of Public Service. It worked, but it was not commercially feasible because coal was so cheap. Moreover, improvements we were building into coal burning plants were producing more and more kilowatts out of every million BTU's of coal. Nevertheless, this was my first personal experience with a consortium of companies to provide financing, and later I used the same method to develop Fermi #1, the first large-scale demonstration model of the fast-breeder nuclear reactor.

Another part of our forward planning took us into the transmission system. Bulk power was carried throughout Public Service on 132 kilovolt (kv) lines. Our studies showed that if we could build interconnections with Pennsylvania Power and Light and Philadelphia Electric Co. we could buy electricity if one of our generating plants was down when needed. Such interconnections would save us from building an additional plant to protect our reserves of energy. We could also sell power to Philadelphia if it needed more than its capacity could produce. As a result a transmission grid was developed between Pennsylvania and New Jersey.

The grid was a particularly important development in another way. The Pennsylvania Railroad decided to convert some of its coal burning locomotives to electricity. I dealt with the road to see whether it would build its own generating plants or take power off the grid. Our economic studies showed that it was cheaper for the Pennsylvania Railroad to buy from us than to construct and operate its own power company. This was my introduction to the concept of a grid, and I gained further experience with the practical operating advantages in Europe during World War II. Thus, I was ready to apply it to The Detroit Edison Company when I joined this organization.

During these years of construction and planning, I was being promoted within the company. I advanced to general superintendent of generation and then to assistant general manager. As my responsibilities increased, my concerns widened. Since there was more generating capacity than needed, emphasis on the Sales Department increased. The market for electric automobiles was wiped out by the internal combustion engine. The electric street car held its own for a few years but gradually lost to the diesel bus. Oil and natural gas preempted the heating and cooking market.

Our planning group began to work with manufacturers of electric appliances to improve efficiency so that their appliances would be economically competitive with those powered by other fuels. Electric refrigerators and washing machines gradually came to dominate the market. At Detroit Edison, Hirshfeld had experimented with the heating units of electric stoves and developed the Electrochef, the first model really competitive with the gas stove. Although I kept in touch with these developments, I began to focus on the critical cost of fuel, and my understanding of economics took on a new dimension. If we could design more efficient turbine generators and run them at their most economical speeds, we could afford to sell electricity at lower rates. Regulatory agencies held us accountable for this development, but as a group of engineers and businessmen we were already dedicated to productivity.

Against the background of my own youth and of the depression years, I saw cheaper power as creating a better way of life for everyone. It also meant a larger and more challenging job for me. There were several opportunities to leave the power industry because I was meeting many businessmen and some of them asked me to join their organizations. I had a particularly attractive offer from one company, and had I accepted I might have become a wealthy man. If I was not getting rich at Public Service, however, I was doing things I wanted. As I look back, I see I was trying to do my part in restoring the balances in the economic system.

Later on, during the war years, after I had met European politicians, government officials and businessmen, I thought about the influence of President Roosevelt. It seemed to me that he was not interested in reform for its own sake and that he had no ideological position. Instead, he was a practicing politician of a very high order—a statesman. He worked to set national goals and to persuade people to support them. The first of these was economic recovery, and in accomplishing this goal he brought government and private industry much closer together than they had been. The second was to unify the country in the support of free political and economic systems, and he was so successful that he won an unprecedented third and fourth term in office. Even when the country was still divided over war or no war, he was able to provide lend-lease support to the allies. It seemed to me that this was the kind of leadership which should be practiced in industry as well as at the national political level.

In 1941, all of my earlier experiences came together for me, and I was sent to Washington to participate in the great government-industry effort of World War II. I had completed my formal education. I had spent nineteen years in Public Service of New Jersey learning my profession. As a part of a regulated industry I had become accustomed to working with

57

government. These factors, when added to my subsequent experiences during the war, shaped my future career.

By the end of the 1930s my personal life had also become settled. When my son was eight years old, he was hit by a truck while riding his bicycle, and despite every effort to save him he died of the injuries. After that my marriage broke up and I lived alone. Among my closest friends were Bill and Gertrude Rippe and their two children. Bill had been a classmate at Cornell, we had joined Public Service at the same time, and during the next thirteen years we shared many of the experiences of that organization. When he was struck by a strep infection, I visited him every day. Just before he died, he asked me to look after his family, and this I promised to do. Over a period of time Gertrude and I talked about the future, and on July 28, 1939, we were married. I adopted Dick and Jane as my own children but felt strongly that they should keep their father's name and perpetuate our memory of him in this public way. It was a joy to have a family again, and I loved them as if they were my own natural children. When they grew up, they both went to Cornell and both married Cornellians. Now we have six grandchildren.

As anyone who reads this volume can tell, Gertrude has been a very understanding as well as a loving wife. She enjoyed travelling, and we went together frequently. Earlier in her life she had taught school, and this gave her a lively interest in many things. Sometimes she would venture on her own if I got occupied in days of technical or financial negotiations, and she greatly extended the scope of my own interests. She met other women very easily whether it was Marguerite, the wife of Ernest Mercier, or Marian Brown, or Gertrude, the wife of Sir Harold Hartley, or Odette de Heem. She also raised the children when I was away at war, and provided me with a home in Detroit around which colleagues and friends and visitors could gather. We have been very happy together.

Editor's note: Mrs. Cisler died November 21, 1975.

CHAPTER IV

THE DETROIT EDISON COMPANY

The men of The Detroit Edison Company were well known in the power industry. They participated in the activities of professional engineering societies, published papers and did research. They pioneered the installation of large efficient machines, and several of their mechanical engineers were as good as could be found. In addition to their technological competencies, their marketing policies dealing with reliability of power and with customer services were models. So when Marshall and Parker asked me to leave Public Service of New Jersey, I decided quickly to do so. It meant giving up over twenty years of associations and losing that much time in retirement benefits to try a new life in a new location. But the opportunities to accomplish something significant in my profession and industry were promising, so at age forty-six I made the change. Thirty months later, when my active military service was completed, I moved to Michigan.

In November, 1945, I came to a city that had passed through forty years of rapid, sometimes turbulent growth and had become mature. Its character as an automotive manufacturing center was set, and the dominant position of Chrysler, Ford and General Motors was established. Now the first generation of management was beginning to pass from the scene.

In my own company, Alex Dow, who had managed the firm since its incorporation in 1903, had retired as president in 1940 and died two years later. Alfred P. Sloan, Jr., who had directed the development of General Motors since the 1920s, retired as chief executive officer in 1946 to become chairman of the board. Walter P. Chrysler remained active in his company until 1940, and Henry Ford, the man who symbolized Detroit more than any other, resigned as president of his company on September 21, 1945.

Henry Ford had worked under Alex Dow for almost ten years as chief engineer of the early power plants in the predecessor Edison Illuminating Company. Then in 1903, The Detroit Edison Company and the Ford Motor Company were both founded. The story of Ford's death in 1947 was reported in all of the newspapers. He had built a small generating plant for his home in Dearborn, but the Easter rains flooded the Rouge River and cut off the electricity. The house was lit with candles, and the nearest workable telephone was at the company engineering laboratories. This man, who had become the worldwide symbol of mass production, passed away during the night with only the supporting services he would have known as a farm boy.

Chief Engineer of Power Plants

They were waiting for me in Detroit even though I reported to work in uniform. It took time to get civilian clothes in those days, and the company was in a hurry to begin planning for peace rather than for war. Many organizational changes had already taken place. Marshall had retired. In the spring of 1944, Prentiss M. Brown, a United States senator and lawyer, had been brought in as chairman and chief executive officer with responsibility for financial and public affairs. Jim Parker, as president and general manager, was the chief administrative officer of the company. As chief engineer of power plants, I reported to Vice President Paul Thompson. Men like Bill Brown, Frank Nitzberg and Arthur Griswold, with whom I had worked in Washington, were now part of my team in Detroit and made me feel welcome. Most of these men had already served the company for twenty years, so there was great continuity in the midst of postwar change. Thompson had been hired in 1913 as experimental engineer at the Delray Power Plant, and it was his reputation as an innovator in the industry that brought him back to Cornell in 1922, when I heard him lecture to the undergraduates. He continued to hold down my position as chief engineer of power plants while I was serving in Europe, and now he had charge of engineering, construction and operations.

Among these experienced men there was considerable debate about how fast the normal civilian economy would be re-established, but it was generally agreed that whether we had a few months or a few years there was still a great deal to do. Because of military priorities no new generators had been installed for five years and only a minimal number of lines added. With the depression before that, there had been below-normal growth for the preceding ten years as well, so in 1946 The Detroit Edison system was very much what it had been almost two decades earlier despite the population and industrial growth of the war years.

A capital intensive industry requires a longer lead time to respond to change in the market than other kinds of manufacturing, and this is especially true of power. Since we have not yet learned how to store large quantities of electricity, it must be produced at the same instant the customer presses a switch to turn on a light or a giant machine. To meet the maximum demand that may last only a few hours a year, capacity must be ready for this peak load. In order to add capability to the system, land must be acquired for power plant sites and transmission corridors. Turbine generators must be installed, coal to fire the steam boilers for twenty-five years needs to be secured, and railroad and water transportation to move the millions of tons burned each year must be obtained. For Detroit, coal had to be carried about 500 miles from Kentucky and West Virginia. So, with all of these long-term factors, decisions had to be made on the basis of the best experience and judgment available. The management's decision was to take the risks and move ahead rapidly with an expansion program.

The company had added less than seventy-five megawatts of capacity since 1941, and now had a capability of generating 1,336 mgw if all its machines were running. But the peak customer demand had increased 32 percent during the war years, so there was very little reserve margin in case of a forced outage. One of my orders from the War Production Board had stopped the installation of a new machine at the Marysville power plant by diverting it to a more critical need. My first job now was to get a seventy-five megawatt turbine generator on the line at Marysville and generating electricity in 1947. We also started immediately to plan the enlargement of Trenton Channel Power Plant that I had visited as a young engineer in the 1920s. And then we began to plan to enlarge the Conner's Creek plant with two more units. From 1949-51, we actually added 400 mgw of capacity and increased the system by almost one-third.

Even at this pace we were able to keep only a year ahead of the surging customer demand that had been held down by wartime restrictions, while Consumers Power, our neighboring company to the west of us in Jackson, Michigan, had very tight reserves. My experiences in Europe had made me particularly aware of the many advantages of interconnected systems, so I felt it was necessary to seek additional ways of strengthening the supply of electricity for the state by increasing the capacity of the tie with Consumers. This interconnection was a 132,000 volt line built in 1928, and the two companies originally had a contract to deliver thirty megawatts in the event of an emergency. When Consumers had to buy fifty megawatts, we turned on water hoses to cool the transformers. In 1946, we strengthened this interconnection to carry the fifty megawatts and in 1949 we doubled its capacity again. Also, in the latter year, we built a second interconnection above Pontiac toward the northern part of our system and, shortly after, a third tie with Consumers Power below the city of Monroe at the south end. When these were completed, the two companies could exchange 350 mgw, which added greatly to their reserve capacities and permitted both firms to meet customer demand under peak conditions.

Before I came to Detroit, it had been company practice to design, engineer and build its own power plants. This was a proud tradition, and it continued through the installation of the new unit at Marysville. As we looked ahead at the construction that would be needed for future growth, however, it became apparent there would be too much for our own work force and that it would be better to contract the jobs. This meant we would need a thousand fewer construction people on our payroll and would require better trained supervisory and professional people to handle the increasing complexity of our business.

From the beginning of the company, the management had taken a somewhat paternalistic attitude toward employees. Salaries and fringe benefits were as good as or better than the standards of the industry. There were paid vacations, sick-leave and insurance, but no pension rights, although President Dow had given some retirement allowances. There were many recreation clubs paid for by the company and the food service was subsidized. As a result, even during the depression years, when Detroit became famous for the growth of industrial unions, the men did not feel the need of organization.

By 1941, however, the early practices and conditions were stated as a contract between the company and the overhead linesmen, cable splicers and some people in the stores department. The men were represented by Local No. 17 of the International Brotherhood of Electrical Workers (IBEW), an AFL union. Over the next several years the skilled and unskilled workers in a number of departments including substation, production, construction, meter reading and building operations also got contracts, and these were merged into a single Local 223 of the Utility Workers Union of America (UWUA), an affiliate of the CIO.

J. Herbert Walker, the vice president of employee relations, who had developed the new management position toward unions, died in 1947, and I was asked to take over his responsibilities in addition to those of chief engineer of power plants. It required considerable time in seminars and training sessions to change some basic attitudes of the supervisors and in many of the men themselves toward collective bargaining, but we had a contract under which differences could be negotiated and grievances resolved. I had worked with Joseph Keenan of IBEW during the War Production Board days and then with Joseph Fisher of UWUA. Fisher had been employed at Consolidated Edison in New York, and his son had attended the Labor and Management School at Cornell. So it was possible to talk with the men and to gain their understanding. George Porter, who knew the rank and file in the power plant on a first-name basis, joined me for a year until I could make him chief engineer. We finally worked out an agreement whereby we completed the additional structure for the two units at Trenton Channel with our CIO local but shifted to an outside contractor to install the machines. Since the contractor had an AFL union, this separation required some pretty delicate negotiations.

During these and subsequent years the management worked hard to maintain the confidence of all employees. As president of the company, Jim Parker gave his personal attention to this matter, attending programs with supervisors and meeting with union representatives. In the decade after 1943, when the first collective bargaining agreement was signed, the average hourly earnings rose only from $1.21 to $1.29 under wartime controls. It rose from $1.50 in 1946 to $2.20 in 1952. And the fringe benefits were then worth 24 percent of payroll.

In 1948, I was named executive vice president of the company. Parker and other executives of the public utilities were interested in the possibilities of atomic energy. He was appointed chairman of the Industrial Advisory Committee by the Atomic Energy Commission, and I, as executive secretary, was responsible for writing the report and recommendations as described in the following chapter. In the same year Paul Hoffman, who was administering the Marshall Plan for the reconstruction of Europe, asked me to continue my wartime contacts for the public utility parts of the plan. This activity is described in Chapter VI. Jim Parker was worried about the company and these additional demands on my attentions, but he felt that Detroit Edison had national and international obligations. At the same time, I was much concerned with the capability of the company and Consumers Power to meet the pent-up load demand as people bought toasters, washing machines, refrigerators and electric ranges, and as the automobile companies stepped up their own production.

As chairman of a committee of the Edison Electric Institute, I organized the first national power survey to learn the needs of the country. A start had been made by Arthur S. Griswold in the Office of War Utilities and another in the Defense Electric Power Administration in which Harold Reasoner had participated. Now both men began to work on a new one. Art Griswold had been in my class at Cornell and, along with Hirshfeld, A. Douglas Campbell and others in the company, enjoyed the confidence of General Electric, Westinghouse and major manufacturers. They opened their order boards to them. The information that trickled into our first effort was so thin that I presented it orally to the Institute. Two years later the reporting organizations understood the categories of information; and data on existing capability, scheduled additions and future plans began flowing in depth. We had established the credibility of the survey. In October, 1949, the first formal report was published and distributed to the industry and the Federal Power Commission.

I remained chairman of the committee for ten years and was succeeded by Art Griswold. He was later succeeded by Harold Reasoner, who continued preparing the semi-annual reports so that The Detroit Edison Company played a leading role in maintaining this important planning instrument for a quarter century. It was a grass roots report with data obtained from all organizations generating electricity—investor-owned utilities and manufacturing companies with power plants as well as federal, municipal and rural systems. One of the lessons I learned from my War Production Board days was that you must get out of Washington if you really want to learn what is going on. The result was an accurate record of the peak load, capability to meet the load and the total kilowatt hour requirements. On the basis of this data, forecasts were made so that manufacturers could schedule their production of turbine generators, transformers, cable and all other essential elements while the operating companies could develop firm plans to expand their systems. The survey put the private industry in a position of strength.

Twenty-five years later, in 1973, a crisis was precipitated in the oil industry through a short-term embargo by the oil producing Arab states. Essential information for long-term decisions was fragmented among so many public agencies and private corporations that a coherent policy could not be formulated. After two years of debate and severe inflationary experience our political leaders are still divided about the future management of energy. The investor-owned utilities were able to avoid the bitterness and suspicion created in the public mind by these events, although they suffered the consequences of this disruptive experience along with everyone else.

In September, 1951, I was named to the board of directors of the company but was told nothing further. Upon the retirement of Jim Parker in December, I succeeded him as president, general manager and chief administrative officer. Prentiss Brown remained as chairman and chief executive officer. He gave me full support on running the internal affairs of the company, and he shared the external affairs. In turn, I was completely loyal to his interests. In this atmosphere of mutual trust, a very close friendship developed which was certainly one of the richest experiences of my life. On July 1, 1954, Prentiss Brown retired as chairman but remained an active member of the board of directors. I became president and chief executive officer. The position of chairman was held open for ten years because of my respect for him. Even after he retired from the board of directors in 1970 and until he died in 1973, a week hardly passed that we did not consult with one another on the phone.

It was about this time that I began to think about the four channels of organization that I have used to keep the details of a wide variety of activities in reasonably good order in my mind.

The Technological Channel

The technological channel developed first. As a student at Cornell, I had trained my thought processes to approach scientific and technical matters in a systematic way. At Detroit Edison, I found myself doing the same thing. The company excelled in load forecasting and constantly tried to make its estimate of the future more accurate by introducing national indices like the Gross National Product and local indicators such as projected growth of the automobile and steel plants. This information gave us enough lead time to build the capacity needed by our customers. We made economic studies of fuels and machines, and the resulting new equipment with higher temperatures and pressures reduced the coal needed to produce one kilowatt hour of electricity from 1.02 to .95 pounds. Such efficiencies saved hundreds of thousands of tons of coal and millions of dollars.

I put Harvey Wagner in charge of the studies of atomic energy to develop a truly significant technological breakthrough that would free Michigan homeowners and job producers from the enormous transportation costs of a coal-based economy. I had not known Harvey before I came to Detroit, but he had been with the company since graduating from The University of Michigan in the mid-twenties. He and his wife Eleanor became warm personal friends of Gertrude and me along with our many earlier acquaintances. This new assignment opened fresh challenges to him, and he made himself an authority in the field.

The company forecasts called for a start on advanced planning for the 1950s, and the quantities of generation that were called for, along with the projected growth of our service area beyond metropolitan Detroit, indicated that we would have to build the first new power plant since Trenton Channel had been started in 1923-24, when three 50-megawatt units were installed. The site for the St. Clair plant, just south of the Marysville plant in the Port Huron area, was acquired in 1948, and by 1954 four new units, each rated by the manufacturer at 150 mgw, had gone commercial. Three years later, another new power plant at River Rouge, in the heavy industrial complex near Dearborn and the Ford Motor Company, was able to generate electricity with two machines of 260 mgw each, at that time the largest in the world. So with this kind of building, along with improvements in transmission, there was enough in the technological area to keep the company busy.

The Commercial Channel

Another channel for information and decision-making involved the commercial aspects of the company. These fell rather naturally into the related groups of marketing, pricing and financing activities.

When I came to Detroit in 1945, Sarah M. Sheridan was still active, and she had become an almost legendary figure in an essentially masculine world of engineers, construction crews, technicians and customer servicemen. She had joined the predecessor company in 1892 as a file clerk, and Alex Dow made her manager of sales for Detroit Edison in 1907. She was then thirty-two years of age. In 1921, Dow appointed her vice president for sales, and she was a rare individual in many ways. She developed the marketing functions of the company during the expansion of the 1920s, the depression of the 1930s and through the war years. She was finely tuned to the needs of the customer, whether industrial or residential. Under her forceful leadership, company people participated in many of the community-related activities where they lived or worked. Special services such as the free exchange of light bulbs, appliance repairs at a cost of parts only, educational programs on cooking and the safety of the housewife were maintained. She was a delightful person, an intelligent and fine looking woman who inspired confidence. Our Miss Sheridan, as she was affectionately called, produced high morale within her department and the public relations result was as good as any in the country.

When I became president, I was determined to maintain this outstanding record. One of the best ways to please the customer was to provide safe, reliable electricity at lower and lower costs. The large new machines were efficient, and a small saving on each pound of the millions of tons of coal we were burning was significant. The interconnections with Consumers Power meant that the company did not have to build and install 350 mgw of capacity in order to meet unusual peak demand or equipment failures. It was an important financial saving. I wanted to extend this practice, and we were able to do so because of the political prestige of Prentiss Brown.

Mr. Brown was an attorney who had spent one term as a United States senator. Michigan was a Republican state in the 1930s, however, and Prentiss Brown was a Democrat. When he was defeated for re-election in 1942, he became price administrator in Washington. This was a sensitive position, which he filled in a way to earn him the respect of consumers and producers alike. He was elected as chairman of the board of Detroit Edison in 1944, the first corporate officer and indeed the first senior man to have come from outside since the earliest days of the company. In addition to my great admiration for his business and legal ability, our families became congenial friends. Marian Brown liked Gertrude. When I speak about him to others, I find that I still tend to call him Mr. Brown.

The existence of the Cold War between the United States and the Soviet Union dominated international politics at the end of the 1940s. When we wanted to create an interconnection to the Toledo Edison system and the Ohio companies, we went to Washington to testify about the potential value of this to national defense. Prentiss Brown was known in the Federal Power Commission and political circles and was able to gain the support of Michigan senators and representatives. The Michigan Public Service Commission also testified for the economics of the proposal and for its own power to represent the public interest.

Harvey Fischer of the law firm of Fischer, Franklin and Ford had been associated with Detroit Edison since 1924, handling day-by-day legal matters in Michigan, and he gave us direction for fifty years. Because the company was incorporated in the State of New York, Paul McQuillen of Sullivan and Cromwell represented our interests with the SEC and the Federal Power Commission. He and Prentiss Brown organized our efforts in Washington while Harvey Fischer and Prentiss coordinated the Michigan support. I drew on my experience in setting up the international grid in Europe. Congress was interested in our position but took no definite action at this time.

We next wanted to strengthen the regional supply of power with Ontario to the east. Since this involved shipping electricity across an international boundary, the FPC claimed jurisdiction. Once more I accompanied Prentiss Brown to Washington, and this time we were successful in getting special congressional action exempting the company from what was felt to be unnecessary federal controls. There was not a single dissenting vote. In August and September, 1953, one line went into service between Detroit and the Canadian city of Windsor across the Detroit River and a second between the Port Huron and Sarnia area across the St. Clair River.

These activities fell basically in the financial rather than the technological channel. The only really serious technical problems to be solved were the control mechanisms on all the interconnecting lines that would prevent a fault in a neighboring system from cascading throughout a whole region.

Such a failure happened in 1965 during the northeast blackout. New York and eastern seaboard power systems were linked to Detroit Edison through Canada, and when a fault occurred in one of the switching stations at Niagara Falls, it tripped out the lines through much of the area for seven hours in the late afternoon and evening. New York City dramatized the plight of an urban industrial center without power. Elevators stopped between floors. There were no electric lights, and the traffic control systems were down. Commuter trains did not run. Many people were stranded all night. Fortunately, there was no panic, and fortunately The Detroit Edison Company protective devices separated us from the trouble. In the middle 1950s we had installed automatic equipment to quarantine a badly faulted system. When the fault was corrected, electricity flowed from Detroit Edison, and Ontario put its hydro power back in, so energy was able to start all the generators turning once again.

While the new plants and interconnections gave customers reliable electric service, they also produced reduction of rates which helped the marketing department become more competitive with natural gas. In 1948, the Michigan Public Service Commission authorized a small increase in the price per kilowatt hour because of the need for additional revenue to finance the large building program. In the next twenty years, however, there were four rate reductions totalling $12.4 million. As a result of the total promotional effort of the company and the growth of population in southeastern Michigan, the number of residential customers increased from 726,000 in 1945 to 911,000 in 1951. Sometime in 1953, the millionth customer began to receive service, and the numbers grew steadily. So did the amount of electricity used by each customer. In 1922, the year I entered the industry, 422 kwh were registered on the meters of the average household for the entire twelve-month period. In 1945, when I came to Detroit, the figure had increased to 1,399 kwh. By 1975 it was up to 6,691 kwh, a sixteen-fold increase during my active career.

To me these figures spelled the growth of the industry to which I had dedicated my professional life, but they also meant much more. They were an index of the increase of productivity and of the enormous economic improvement in people's lives. The kinds of hard physical labor and financial insecurity that prevailed in my youth were growing less common. Electricity was doing more of the work around the home, on the farm and in the factory. As a result our customers had more leisure and more discretionary income whether they represented a family with children or the General Motors Corporation. Although I did not fully recognize it for another decade, we were preparing the basis for a whole new life style.

The well-being of the shareholder also lay in this commercial channel. In 1945, there were more than 6.3 million shares each earning a dividend of $1.20. By 1952, there were more than 10 million and the dividend was $1.40. This was increased steadily until 1961, when it reached $2.20 a share and then split two for one. In 1966, the dividend was again at $1.40, so its value had doubled.

All these matters related to the commercial concerns of the company, and I kept them sorted out in my mind in this category. I also began to organize my thinking about the social channel in the early 1950s and came to understand that there were major forces working on me and on The Detroit Edison Company.

The Social Channel

I will describe these forces at greater length in subsequent chapters, but I want to comment on the broad implications of them here. In the postwar years the universities were teaching systems engineering, and every year the systems grew larger. Gradually, we came to understand in new ways how all technological, commercial and social systems are interrelated.

I was still very active in the work of the Marshall Plan although I did not leave the country between 1950 and 1953 because I was so busy in Michigan. But the Cold War between the United States and the Soviet Union, along with the threat of the atomic bomb, influenced the activity of the company. Defense contracts kept Detroit industry busy. The Korean conflict dominated foreign policy thinking for several years and took a number of company men back into military service. As a whole, the world appeared to think only incidentally about peace. Nationalism was rising, and there were conflicts within the developing countries. I worked hard for the peaceful uses of the atom and for adequate sources of other kinds of energy around the world, believing that adequate supplies would correct some of the major imbalances and would lead to the improvement of life everywhere.

Detroit Edison was also influenced by national events. Although we thought of ourselves as a local company tied down to our franchise area in southeastern Michigan, we had begun to ship electricity outside the state, and we were certainly affected by national and international developments. The country had developed a mature industrial economy, and people enjoyed a level of affluence never before experienced. Work was still important, but national productivity had risen to the point where leisure was beginning to compete for a person's time. These were new conditions for millions of people, and they were very pleasant. They proved that our systems were working.

Social changes taking place within our own service area also had to be followed with care. The population grew, and so did the number of political structures of the region. A second and third ring of suburbs began to surround the core city. These were just bedroom communities, however, and the commuting traffic choked the streets until a system of expressways was built. All this was unplanned growth, and Detroit Edison distribution lines had to follow it. The pressures were not yet great enough to force a change in our institutions or even in our way of thinking about them, but they were building.

I became aware of all these changes, but I did not see their direction at first nor perceive their significance. They were all drawn together in the social channel of my thinking, and I was able to keep track of them. Gradually I became able to think systematically about them.

The People Channel

Finally, I began to organize my thoughts about management in the channel of people. With the improved technology we had begun to increase productivity to a significant degree. In the first ten years after I came to Detroit the production of kilowatt hours increased two and a half times while the employment rose only 13 percent. Some of this improvement was passed along to employees each year through negotiations with the unions. Local 223 of the UWUA called out its members in 1952, and three years later Local 17 of IBEW struck the company. On the whole, however, there were good employee relations. The

71

average hourly wage rose steadily, doubling in a ten-year period, and the fringe benefits were among the best in the area. I think it is fair to say that for the most part people were proud to work for The Detroit Edison Company.

Salaries and working conditions for our professional people also concerned me. I believed very strongly that it was important for men to be really interested in their work. I encouraged them to join professional organizations, to become members of the working committees, to do research and write papers, and I saw that their expenses were covered. In an advanced industrial organization, there is no substitute for a creative environment and intellectual activity.

The year after I became president, I started to open up the organization. The title of manager had been restricted to one man. Dow was the first general manager, then Marshall, Parker and finally myself. The general manager had responsibility for all the daily operations of the company, but I wanted to delegate authority for decision-making to various functions. It was possible to do this by discontinuing the title of general manager, thus pushing the top level up and inserting a band of five new managers for construction, engineering, operations, purchases and sales. A few months later, a manager of employee relations and a manager of union relations were added. Over the next five years nine vice presidents were created, and a number of the managers were promoted to this position. We also created a special category of middle and upper management who were paid at a special rate because their responsibilities required them to be available twenty-four hours a day and who frequently worked long hours. It became an honor to be placed on the Monthly Roll, and people took pride in their jobs. Largely as a result of these changes, I was awarded the Henry Lawrence Gantt Gold Medal in 1955 for management.

As I see it, the primary task of management is to understand the elements of the four channels of organization. This is followed by the job of setting goals, winning acceptance of them and motivating people to meet them. The whole process is called leadership, and it functions in military and political life as well as in corporate affairs. It happens in our personal lives as well, so everyone is a manager in some measure.

Throughout the first seventy-five years of this century the electric power industry has been operated primarily by men. Vice President Sarah Sheridan, who was a principal corporate officer of Detroit Edison for thirty years, was a notable exception. Nonetheless, women played an important and complex role.

I always saw to it that wives were invited along with their husbands to conventions and major meetings, and this practice was common throughout the industry. In addition there were the normal friendships that grew between families of people who worked together for many years or who lived in the same neighborhood. Gertrude and I had friends we saw quite regularly and who broadened the business experience for me into a rounded social life. Since the company was limited to its own service area, it did not have to move people around the country as was common in many corporations. The jobs were stable. The average length of employment was over twenty years, a considerable number of people worked forty to fifty years with Detroit Edison, and there were second and even third generation employees from the same family. After World War II, the management turned away from the earlier paternalism, but there still remained a strong sense of belonging and of being responsible members of a kind of extended family. The role of women in all this has never been adequately evaluated, but it was a most important factor in providing stability and reliability in the industry.

The New Phase

In 1962, I was sixty-five years old, and the board of directors asked me to continue as president. Although it looked in the late 1950s as if our construction had caught up with the postwar economic expansion, it was now clear we would need a major new program to keep up with the demand for electricity. The automobile industry was continuing to expand operations in southeastern Michigan. Our forecasts showed a potential rate of growth of almost 10 percent a year, and although there was skepticism and some disbelief in these figures, there was no doubt we would have to double the size of the generating capacity. Opinions differed only in whether this would have to be done in seven or ten years. Under these conditions I agreed to serve for ten years longer provided I was physically and mentally able to do so. Each of the directors told me personally that he wanted me to stay.

Serious planning for the future expansion was now increased. We had agreements for cooperation with Consumers Power extending over many years, and we worked out in detail how we would schedule future construction of power plants and transmission lines which constituted the greatest investment. We began to build a central dispatching center in which computers called the machines of either company on the line according to the most economical unit that was available. On December 22, I went to Jackson to sign with A.H. Aymond, chairman of Consumers Power, an agreement that the bulk power systems would be developed as if there were single ownership.

As a result, Detroit Edison proceeded immediately with the design and construction of No. 9 at Trenton Channel and No. 7 at St. Clair power plants. Each of these machines was capable of producing more than 500 mgw, and when they were completed as planned in 1968 and 1969, they added more than 25 percent to the capacity of the company. We also began to plan the Monroe power plant for the early 1970s and agreed to install four 800 mgw machines.

At that time, I was convinced that nuclear reactors had not yet demonstrated complete reliability for short-term needs while coal had. Consumers agreed that they could proceed with a nuclear plant while Detroit Edison first went along the more tested route. Once the decision was made and approved by Consumers Power, the next major question was the supply of fuel. Earlier in the 1950s, we had studied the feasibility of building a pipeline across Ohio and moving coal through it in a liquid slurry. Then the railroads gave us a very favorable freight rate for a long-term guarantee, and the Company saved $40 million over a ten-year period. This time I asked Consolidation Coal Company and the Pennsylvania and the New York Central railroads to join in a study, and we negotiated our final contracts by putting everything on the table.

Detroit Edison invested $8 million in the mines and contracted for 100 million tons of coal over twenty-five years. Each of the big machines alone required two million tons a year. We also agreed to purchase five locomotives and 150 cars for a unit train operation. The design of this train called for more technological advances. Each car was capable of carrying 125 tons and was made of aluminum to lighten the nonproductive weight. And it was built with a rotating coupling so that it could be tipped and emptied at Monroe while still moving around the loop track at the plant. In this way the turn-around time was kept to a minimum and, with a seventy-two hour schedule for the 700 mile trip, the train was kept rolling almost continuously. George H. Love of Consolidation Coal said, "I'll figure 16⅔ cents per million BTU for the coal," and Stuart T. Saunders of the Pennsylvania Railroad said he could deliver it to Monroe for a total of 22½ cents per million BTU. The contract was subject to labor and escalation factors, but it produced for Michigan by far the cheapest imported energy up to this time. In fact, the Monroe installation was a complete economic system for the production of electricity such as I had considered for Europe under the Marshall Plan, and hoped to help build in India in the Damodar Valley west of Calcutta.

While all of this was going forward, we also began to plan for a large nuclear plant to add to Fermi #1, the demonstration model of the fast breeder reactor being constructed just north of Monroe. I was convinced that nuclear power was the answer for the next fifty years beginning in the 1980s. We already owned the land and had gone through the safety hearings for Fermi #1, and so Fermi #2 came into focus. At first we considered an 800 mgw boiling water reactor, but subsequent investigation showed we could move to 1150 mgw at little increase in cost. I felt we had the men for the job. Detroit Edison was the most knowledgeable nuclear company in the country, and by this later date we would have enough experience to solve any operating problems that might arise.

No matter how fast we worked during the early 1960s, however, the growing demand for more electricity kept pace. Our forecasts were accurate, and there were no downturns in the economy to give us a little more time. We had a plant site at Harbor Beach at the northern tip of our service area, and a small plant could be built there in a couple of years. It came on the line with 114 mgw just in time to meet the peak load of the summer of 1968. We also added an oil-fired boiler to supply steam to the generator at Fermi #1 when the nuclear reactor was between tests. This provided another sixty-two megawatts, and because we did not need to build an entire plant, it was ready in the winter of 1967. By this time, both General Electric and Westinghouse were manufacturing small diesel generators that, although still very large by World War II standards and the floating power plants, could be assembled quickly. These were called peakers because they were designed to run only a few hundred hours a year to meet the maximum demand. By the summer of 1967, we had units installed in four different locations to help balance the system, and they gave us another 276 mgw capacity. Even then, oil was much more expensive than coal, so our fuel bills rose as the peakers had to be used more and more hours to meet the base load.

Michigan had generated a small amount of electricity from rivers in the early days of the century, but it had essentially no hydro power. I had studied this question thirty years earlier in New Jersey. Now the two companies decided to build a pumped storage project at Ludington on the shores of Lake Michigan. This was completed in only three years and proved very valuable in meeting peak demands on a daily basis. Water was pumped up the bluffs at night when there was surplus power and allowed to run down through the turbines during the day when demand was high. It could generate 2400 mgw for an eight-hour period. This electricity cost less than a fifth as much as the average produced by the thermal plants, and as the price of fuel continued to rise it became even more economical.

In 1964, I recommended that the board of directors name Donald F. Kigar as president and chief operating officer. I assumed the title that Prentiss Brown had vacated ten years earlier and became chairman and chief executive officer. Kigar had experience in construction, purchasing and union relations and had a firm sense of the operating side of the company. He also knew the great body of employees and had good rapport with them. We worked together as a team, and all went well until 1967. In that year the Michigan Public Service Commission ordered another rate reduction, and with our rapidly growing need for capital to finance the very large construction program, this action could hardly have been more ill-advised. It reduced the company's earnings at the very time we needed to increase them.

Just before the annual meeting in the spring of 1967, Don Kigar decided, for family reasons, to take early retirement. I was disappointed at losing my successor, but the company had many experienced administrators. I turned to Edwin O. George, who was one of the men promoted when the management of the company was opened up in the early 1950s, and he later became senior vice president and then executive vice president of marketing in the middle 1960s. His long association through the marketing department with our customers and community affairs was particularly valuable at this time of extraordinary social change. It was a good selection.

76

The civil rights movement was growing, and Detroit was beginning to seek new ways of accommodating the addition of poor migrants from the South. The plundering and rioting that shook the city in 1967 proved we were moving too slowly. But change was going on throughout our service area, and the patterns of affluence were popularly described as the new life style in every stratum of society. Young men grew beards and long hair reminiscent of the simpler days of their great grandfathers. Young women became liberated from old customs and laws. It was as if two centuries of industrial development had reached fruition and people in the United States were reveling in the harvest of their efforts.

In the preceding five years, the market had begun to shift to meet the new pattern. The use of electricity had become so nearly universal that it was one of the most sensitive of social and economic indicators, and what happened "out there" in the market was felt instantly "in here" at the company. The flick of a switch delivered electricity where the work was to be done and, although many people thought it was in the walls of their homes, this simple action called on generators miles away. In the factories the switch used to be turned on by hand. Now the flow of electric energy was controlled by delicate sensing devices, themselves run by electricity, that could regulate the amount precisely to conform to human wishes. The tremendous surge of power that flowed along the cable when an electric blast furnace was turned on to produce steel was awe-inspiring. Of more extensive impact, however, was the development of air conditioning in factories and homes. By 1967, the peak demand for electricity shifted in the Detroit Edison's service area from the long nights and cold weather of the Christmas season to the long days and heat of summer. This statistical fact, apparently so simple and of minor importance, actually signalled major changes going on in the daily lives of our customers.

The housing industry underwent a revolution during this period. Individual homes continued to be built, and they grew larger, requiring more land. In addition, more people began to own two larger houses, one in the urban area related to jobs and one in the rural area related to recreation. A house trailer or a boat might substitute for the second home, but the resulting increased use of energy was the same. Manufactured housing in the form of mobile homes began to supply an even larger share of the market, and so far as the company was concerned, this was instant housing. Where the customer representative formerly had six months' notice that a residence was being built, he now began to receive calls from a housewife announcing that she and her children had driven in yesterday, her husband was on his job today, and where was the electricity!

Traditional houses were built to keep warm air inside. This had been important in my youth when we had hand-fired furnaces and the limited supply of wood or coal was relatively expensive. The new-style home began to be oriented to the outside. People added patios with cooking appliances and outdoor lighting, and swimming pools with pumps and heating units, and they installed electric air conditioning inside for year-round climate control. So there were not only more housing units to serve, but each one used more energy to provide the kind of living people wanted and could afford.

As the population moved farther out from the old central business district, shopping centers began to replace individual stores, and the subdivisions that rapidly grew around them, including high-rise apartment buildings, grew into cities within a ten-year period. No one had experience with the rapidity of this kind of growth. Industry and commerce moved toward the open, cheaper land of the suburbs, which now changed from bedroom to balanced communities. Like the new houses, these centers were much more open than the older department stores that had developed fifty years earlier. They used great quantities of electricity for warming and cooling the air and pumping it through the buildings.

If mass transportation between towns was less efficient than it had been earlier, electricity was now being used increasingly for the indoor movement of goods and people. Detroit Edison had 100 percent of the vertical transportation market of elevators and escalators. It dominated inventory handling within buildings that frequently occupied several acres of land. And peoplemovers that used electricity for power were being introduced for horizontal movement.

With increasing affluence, the health industry expanded, demanding sophisticated research and controls that could run only on electricity. During my own lifetime, some twenty years have been added to the average life expectancy. Tourism became one of the largest industries of Michigan as leisure increased, and along with the movement of people there were motels and restaurants that added still another part-time residence for a family. One piece of the entertainment market moved back into the home. When television replaced the movies, an individual no longer joined a thousand others for a two-hour motion picture show a week. Instead, a thousand TV sets were purchased and ran for hours every day. Entertainment blended with education and communication. Computers and information satellites were added so that in the end Americans had much more information than ever before and the human mind was stretched as it never had been. The day came when we actually watched a man walking on the surface of the moon.

In place of familiar words like city and municipal, people began to use urban and metropolitan to describe the built-up area where they lived. In southeastern Michigan, the boundaries between urban and rural areas began to disappear. Farmers enjoyed the amenities of cities as the new level of industrialization began to blur many former distinctions between town and country. The term agri-industry replaced agriculture. With new machines and seeds and livestock, the age-old fear of hunger had been pushed far to the background of the United States. The current problem was obesity rather than starvation.

All these changes were reflected in the company's load, its construction schedules, the financing of new facilities and the supply and maintenance of the system. I engaged Doxiadis Associates to make a five-year study of what was happening in our service area, and we began to acquire future plant sites much farther from the present population centers than anyone had thought of before. In Europe and Japan following World War II, and in the developing countries in the 1960s and 1970s, there were shortages of electricity. But the American people had never experienced such scarcity. The power industry had always been managed so that when people flicked a switch, energy appeared immediately. And it was almost completely reliable. Winter ice sometimes broke transmission lines, hurricanes blew trees over distribution wires or tore up poles and towers, lightning knocked circuits out, and generating equipment even broke down. But the men of Detroit Edison had built the system so well that power failure was down to 1 percent. Through many years of experience customers had come to expect that they could have as much electricity as they chose to purchase and at any time they wanted it.

Ed George understood these social changes and worked hard

to get the company in position to meet them. When he retired in 1970, I recommended that William G. Meese be named president and chief operating officer. Everyone in the company thought highly of him. He was a sound administrator, who had come up through our engineering research department to executive vice president for production. In 1971, he was named president and chief executive officer and became fully responsible for managing the company. I remained chairman of the board.

Bill Meese showed his leadership in the industry by taking a principal role in formulating a national research policy. A few individual power companies maintained their own research capabilities, and Detroit Edison had the largest. In general, however, the industry relied on the manufacturers to develop the new products that enabled us to do a better job for less money. Whatever countrywide program existed was centered in the Edison Electric Institute with a budget that had been gradually increased to $3.5 million a year. Bill Meese joined with a few other executives to establish an independent Electric Power Research Institute with a plan to develop an annual budget of about $350 million. When this was added to what the suppliers were spending and what the federal government was beginning to consider, it was an amount commensurate with the needs for new sources of fuel and a very sophisticated technology.

By this time, however, the financial difficulties of acquiring sufficient capital for the large building program had begun to overshadow all the technical, commercial and social issues. Until the early 1960s, economies of scale were passed on to customers in the form of rate reductions, but by 1967 cost increases had become serious. The change came so quickly that many people were unprepared to respond. In recent years, the public service commissioners were always well-trained lawyers who knew very little or nothing about the economics of the industry before assuming their responsibilities. Although they tried to familiarize themselves, they were political appointees, and this was a primary force affecting their decisions. They had to rely for information on a small staff and on an operating procedure that had been satisfactory in a period of slow growth but was quite unrealistic in this inflationary time. Detroit Edison and Consumers Power finally persuaded the legislature to permit the companies to pay for a larger staff.

But the times overtook them. The government bureaucracy

was simply too remote from the people and the problems to understand the trends. In 1967, we asked for a $30 million increase in rates, and the Public Service Commission granted $6.5 million the following year. From then on we returned regularly, but the results were always too little and too late. They were subject to political influence to a degree that had not been applied for thirty years.

Since the United States is going through another of its periodic cycles of increased government control of business, a consideration of the experience of the power industry is enlightening. As organizations, public utilities occupy a middle position between public agencies and private corporations. The former are subject to political control and financed out of tax dollars. The latter are subject to market competition and funded out of profits. Like private corporations the utilities are investor-owned, they raise their money in the private market, and they pay taxes. But they are also publicly regulated and lack authority over their pricing policies. So they have been easy targets for politicians who want the popular appeal of lower prices and who are not responsible for long-range costs.

Electric power is a capital-intensive industry and requires an investment of four times more money in plant and equipment than most other manufacturing companies for each dollar of revenues. Meeting our projected needs for electricity in southeastern Michigan meant doubling the size of the system between 1966-75, and this required the company to raise more capital than it had done in the prior sixty years. The alternative to this was not cheaper electricity but economic slowdown. Energy studies conducted all over the world made one thing clear. A society must provide about one-eighth of all its capital needs to make energy available for a modern industrial state. If it does not, the other seven-eighths of industrial activity, including food, health, housing, transportation and the jobs associated with them, will grow less and less productive. Every aspect of life will be affected and the real quality of living will decline.

By 1975, the interest rate on utility bonds had reached 12 percent, just when the company needed to spend over $500 million a year on expansion. At the same time, the rate of earnings allowed by the regulative process was less than 7 percent. The company had reached the unhappy position expressed so well by Mr. Micawber when David Copperfield visited him in debtor's prison: "He solemnly conjured me, I

81

remember, to take warning by his fate; and to observe that if a man had twenty pounds a year for his income and spent nineteen pounds, nineteen shillings and sixpence, he would be happy; but if he spent twenty pounds one he would be miserable."

When the oil embargo and the energy crisis struck the country in 1973, the company had insufficient control over its income to respond adequately. Unable any longer to borrow money at interest rates that could be paid out of earnings, and with the price of common stock about one-half its book value, there was no option but to defer construction. This was true in other states as well as Michigan. With us, however, it meant deferring nuclear plants, the one source of energy that made economic and environmental sense. The state experienced zero growth of electric energy for the first time since the great depression. Much depends on how long the financial imbalance lasts, and the many consequences remain to be unfolded.

In 1975, I retired as chairman of the board after a full and satisfying career with the company. Bill Meese moved up to chairman and remained chief executive officer. John R. Hamann, who had been a power plant superintendent and later executive vice president for operations even as I had been a generation earlier, was named president and chief operating officer. He knew the men and women of the company and through his long experience had earned their confidence.

During times of great stress it is always difficult to know whether fundamental changes are being introduced that will radically alter the shape of things to come. In my half-century in the industry, however, the curve of electricity usage has gone up. Even though the depression and the war flattened the growth rate, these plateaus look now like minor variations in a long rising slope. Everything I can see in the basic technological and economic forces indicates that the generating capacity of the company must soon double and then double again. I feel very confident that Detroit Edison has people with the ability to manage this growth.

CHAPTER V

ATOMIC ENERGY

Although the industrial revolution produced spectacular technological achievements in the nineteenth century, American businessmen in the twentieth century have enjoyed more stimulation from research and growth than any others in history. I have always found pleasure in building bigger power plants, and in my lifetime the size of turbine generators has increased more than twenty-five times. I have also enjoyed building larger organizations and systems to distribute electricity more effectively. But the discovery of atomic energy and the research that led to its development presented an opportunity to be involved with something never before accomplished.

The metal uranium, as a source of primary energy for a controlled chain reaction, had not been used up to this time. In the early years of the century, scientists like Madame Curie experimented with radium and other radioactive materials and applied them principally to medicine. But the conversion of uranium for electric energy required the invention of a new technology. From the earliest records, people used organic fuels such as wood and animal oils to heat their shelters and cook their food. Within the past two hundred years we have used fossil fuels like coal, oil or natural gas, and we converted them into heat through the process of combustion just the way we burned wood. But uranium produced heat through the process of fission, in which energy was created by splitting the atom.

As I look back today, I realize that I acquired my introduction to atomic energy in casual ways. In the late 1930s, general announcements came out of Germany on the research of Hahn and Strassmann. These scientists indicated that there were great sources of energy locked within the uranium atom. I did not seek out this information, but because I was employed in the power industry, it came to my attention. As was typical of the day, the employee magazine of The Detroit Edison Company, *Synchroscope,* published an account in January, 1941, entitled "U 235—What Is It?" My own interest was hardly deeper than this popular level.

On December 2, 1942, however, the first controlled chain reaction was produced by Enrico Fermi and his associates under the football stadium at Stagg Field in Chicago. I knew of Fermi by reputation as one of the brilliant scientists who had taken sanctuary in this country from the Nazi and Fascist oppression in Europe. Later, my activities with the War Production Board brought me in contact with the Manhattan Engineer District Project being conducted at Oak Ridge, Tennessee, and Hanford, Washington, and once again I became aware that work was going forward in atomic fission.

This research was being done under the command of General Leslie R. Groves, and it was highly classified. No one asked more questions than were needed, but these projects were competing with others such as synthetic rubber and high octane gasoline, and all military needs were competing with the civilian economy. Oak Ridge required electricity of varying frequencies, as well as copper, steel and plant building material. This need in turn required a rescheduling of equipment and resources on the order boards of various manufacturers, a matter within my area of responsibility. First, a unit ordered for a municipal power system in Indiana, and then three units of 25,000 kilowatts each for a copper project in Utah were shifted to Oak Ridge. But these decisions were just a few among the many dealing with power for conventional uses, which appeared more important to me at the time.

Then, in 1945, the silence was broken when two atomic bombs were exploded over Hiroshima and Nagasaki in Japan. The energy potential of uranium became public knowledge, and the power industry was immediately interested in it as a possible source of electricity. The Committee on Power Generation of the Association of Edison Illuminating Companies tried to learn more about this new source of primary energy, but most of the information they needed was classified as a military secret. Even when the war ended, the international feelings of suspicion forced people to consider the strategic advantage that lay in the control of atomic energy.

It was not until 1945, when the first Atomic Energy Act—the McMahon Act—was passed, that civilians were permitted to consider how this new energy source might be developed to produce electric power for the economy. David Lilienthal, who directed the development of the largest public power system, the Tennessee Valley Authority, was made the first chairman of the Atomic Energy Commission (AEC). Although the act declared its objective was to promote world peace and to improve the standard of living and general welfare of people, the commission found that military security limited the civilian use of the great store of technical information though control had passed from General Groves and the Manhattan Engineer District.

The commissioners soon realized that industrial participation was necessary if this new source of energy was to be made available to people in general. I had been associated with Lilienthal in the War Production Board, and since I had recently returned from the Allied Military Government in Europe, he asked me to serve as a consultant to recommend regulations and controls for the overseas interests of the AEC. In April of 1947, I was given a Q clearance for security purposes, and in October, Lilienthal addressed the Economic Club of Detroit on the dilemma between the military need for secrecy and the civilian need for open business competition. At a press conference after the meeting, he announced the formation of the Industrial Advisory Board to help the commission.

Jim Parker was named chairman, and I was the executive secretary. The board included scientists and experienced utility executives as well as other industrialists. These men had helped keep the civilian economy functioning during the war while adding the productive capacity needed by the military. They understood how important this balance was and that the public had to have access to the technology of atomic energy if the civilian side was to remain strong. During the next several years national security had primary consideration, and this required businessmen to work with military men as had been done during the war.

For the next year, in addition to my daily operating responsibilities with The Detroit Edison Company, I worked with people I knew in the government, military and industry to find ways of resolving the many complex issues. The Industrial Advisory Board was interested in the use of atomic energy for medicine, agriculture and industrial production. Because of the legal complexities, Herbert Marks, the first general counsel of

AEC, joined us. He had been in the War Production Board under John Lord O'Brien and before that with TVA, and earlier still he had been associated with Supreme Court Justice Felix Frankfurter and with Judge Learned Hand in New York. It would have been difficult to find a more able representative of government interests. Marks and I wrote the report on weekends and in evenings. The Industrial Advisory Board recommended that private ownership and initiative should take its appropriate place along with the military. Conditions were suggested under which fissionable fuel would be made available and permission would be granted to manufacture the handling equipment. The report was printed, and it accelerated the plans that were in the minds of many people, both in government and out, to put this energy to work in peaceful uses. It also laid the basis for amending the Atomic Energy Act.

The Peaceful Uses of the Atom

Almost exactly two years later, in December, 1950, the AEC requested more plans on how to advance the peaceful applications of atomic energy. The Commission and the military had studied its use to propel submarines and aircraft carriers, and these applications were being pressed under the leadership of Admiral Hyman Rickover. He had gathered a number of experienced engineers and scientists and had begun to train younger men. The submarines *Seawolf* and *Nautilus* were being readied to demonstrate the feasibility of using a nuclear reactor instead of oil to generate electricity to run the vessels. The industrial companies, without technical information or financial resources organized to develop the technology, were still at a severe disadvantage compared to the military organization. Nonetheless, people persevered.

While uranium (U 238) was itself an inert metal that was found in widely scattered parts of the earth, it was associated with an active isotope, U 235, which was subject to fission. Under suitable conditions, an atom of U 235 would split into two lighter atoms and in the process would release two or three particles called neutrons. These particles would collide with other atoms of U 235 and, under controlled conditions, the fission became continuous and a nuclear chain reaction was started. The combined mass of the lighter atoms and neutrons did not quite equal the original fuel atom because the difference had been converted into energy in accordance with Einstein's formula $E = mc^2$, where E = energy, m = mass and c = velocity of light.

The question was how best to use the chain reaction to serve the peaceful needs for energy, and two different approaches seemed feasible. One was to build a thermal type reactor using water as a moderator and coolant to slow down the neutrons and remove the heat generated during the process of fission. Natural uranium was enriched with about 2-5 percent of U 235 and formed into pellets about one-half inch in diameter. The pellets were then inserted in a 12-foot metal tube, and a bundle of these tubes was fastened together to form the fuel core that was suspended in the water. The water could be in the condition of steam, pressurized water or a special form called heavy water.

The neutrons released in fission travel at very high speeds, and it was desirable to slow them down or moderate their kinetic energy, converting it into heat energy. Water, which did not readily absorb neutrons, was a good moderator and also cooled the uranium core. In the pressurized water reactor, the heat of the coolant was transferred to another loop of water, which was shielded from radioactive particles and which could be used to produce steam to drive the turbine. In the boiling water reactor, the coolant, although radioactive, was sent in the form of steam directly to the turbine generator.

The second approach was similar in some ways to that developed for the *Seawolf*. In this type of reactor, the active isotope U 235 was surrounded by a blanket of U 238. Even after it was used up in a water reactor, depleted uranium was fertile and could be bred or transmuted into fissionable material. Under proper conditions, when the neutrons from the U 235 struck the U 238 atoms, they produced a nuclear reaction and turned the uranium into a different and man-made element, plutonium, which was also fissionable. A reactor which permitted the neutrons to move rapidly was called a fast breeder because it produced a new offspring, plutonium, from inactive isotopes of natural uranium and in a quantity greater than the fissionable material consumed. The light water reactors produced about six atoms of plutonium for every ten atoms of U 235 that they burned up, but the fast breeder produced fourteen atoms of plutonium for every ten atoms it consumed. Thus, the world's supply of fissionable material could be greatly increased in the process.

This was a very important discovery, since active U 235 normally constitutes only 0.7 percent of uranium, and we have even to this day found only limited amounts of the metal. By breeding plutonium, however, it was possible to increase the fissionable output from every pound of uranium by more than eighty times, from 0.7 percent to about 60 percent of the uranium atoms. If there were enough natural uranium in the world to last twenty-five years in a light water reactor, its life would be extended to 750 years in a breeder.

In the breeder reactor, the heat from the active core and the surrounding blanket was absorbed by liquid sodium, which then transferred it to a loop of water that was shielded from the radioactivity. This water in turn was heated enough to produce steam to drive the turbine generators. The sodium coolant had several advantages over water. It did not significantly slow down the neutrons as did water. Furthermore, because it did not capture as many neutrons, it helped make breeding possible. And it did not have to be used under high pressure, which made it simpler to contain. On the other hand, sodium reacted chemically with many elements, and special steel alloys had to be used in the manufacture of pumps and the containing vessel. Precautions had to be observed to prevent any leaks in the system and especially to avoid contact with water.

It can be seen from even this brief description that a great deal of scientific information about atomic fission was available by 1950. What was not known were the technological, financial and legal requirements that would translate this knowledge into energy useful to the civilian economy. The political climate in which this advance had to be made in the United States was very mixed.

The international situation was also most complicated at this time. The Cold War had reached threatening proportions in Europe. The Chinese Nationalist government was being pressed back on the mainland and had to go to Taiwan. In 1949, the U.S.S.R. exploded an atomic bomb several years ahead of the time when nuclear scientists in this country thought it would have acquired the knowledge and technology to do so. There were charges of communist spies obtaining atomic secrets, and there was the Korean War. When people all over the world thought of nuclear energy, they had to consider the existence of the atomic bomb.

As early as 1941, however, I had worked with people in the Roosevelt administration at the War Production Board and as a commissioned officer in the Army during the invasion of Europe and subsequently with the Allied Military Government.

I had also worked in the Truman administration under the Marshall Plan. Thus, I knew and respected many people in the government and, in turn, I was acceptable to them. As a result, I found myself able to bring bureaucrats, businessmen, engineers, politicians and scientists together and to change a climate of hostility and suspicion into one of willingness to look seriously at energy as a national problem.

Before long, people were ready to move. Commonwealth Edison (Chicago) and the Bechtel Corporation explored the boiling water reactor. Pacific Gas and Electric (San Francisco) and the General Electric Company went together in another study of reactors. Detroit Edison and the Dow Chemical Company also decided to look into the use of uranium to generate electric power. While the entire United States had vast, though finite, quantities of fossil fuel, the state of Michigan had almost none within its borders. Coal, gas and oil were imported over very long distances with consequent add-on costs to local industry and consumer. It seemed only prudent from the viewpoint of both the consuming public and the producing company to look into a new source that would be free from polluting effects of fossil fuels and not subject to severe penalties for transportation.

In December, 1951, we submitted our report to the AEC and asked permission to continue our work in the breeder field with a fast neutron reactor. The following March we signed a contract with the commission.

My activity began with a feasibility study. A nuclear power development department was created within The Detroit Edison Company under Harvey Wagner. It soon became apparent that nuclear energy was going to require a whole new technology and could be the largest research and development project ever undertaken by private business up to this time. The concept of R & D was just beginning to be understood by businessmen and accepted as a legitimate use of risk capital. The word research still called to the minds of many people the picture of a lone scientist experimenting in a laboratory or an eccentric inventor tinkering in his workshop. As a result of the extensive collaboration of businessmen and scientists during the war, however, a new attitude toward knowledge and a new respect for it came into being. In order to insure an adequate financial base, I asked a number of corporations to come in, and by June, 1953, twenty-six companies agreed to contribute support. One additional task remained. To be able to own or lease nuclear fuel held in government stockpiles for purely military purposes and to receive permission to build, own and operate nuclear power facilities, it was necessary to amend the Atomic Energy Act.

All of these conditions made it desirable to bring about change in the intense fear and suspicion of the day, and I participated in a number of activities to bring about greater understanding. At the suggestion of Dr. T. Keith Glennan, a nuclear scientist and former member of the AEC, then president of Case Institute of Technology in Cleveland and later administrator of the National Aeronautics and Space Administration (NASA), I helped convene a meeting in Detroit of some 200 interested people. The participants recommended the formation of the Atomic Industrial Forum, a national organization to inform people about nuclear energy, and I was elected chairman. We raised money for educational programs in schools and universities, for informative articles in magazines and newspapers, and for appearances before congressional and administrative groups. The forum's effort to answer questions about nuclear energy is still continuing.

President Eisenhower, of course, was very knowledgeable about the military applications of the atom, but he also encouraged the civilian uses. He addressed the United Nations, saying that the United States was determined "to help solve the fearful atomic dilemma—to devote its entire heart and mind to finding the way by which the miraculous inventiveness of man shall not be dedicated to his death but consecreated to his life." He then helped establish a private organization, the Fund for Peaceful Atomic Development, to advance the international development of this new source of energy. I had worked with the general both during the war and later in the reconstruction of Europe, and he asked me to be president of this organization. He had political problems with other nations because the United States had a western monopoly of nuclear armaments. A board of businessmen, labor leaders and scientists was organized, and I visited thirteen countries in Europe to talk with key people. What they needed first was all the published research material that could be declassified in order to update their knowledge. On my return I conferred with Lewis Strauss, who was chairman of the Atomic Energy Commission, and together we outlined a proposal for an International Atomic

Energy Agency to be set up under the United Nations. The first conference on the Peaceful Uses of Atomic Energy was held in Geneva in 1955, and thirty-eight nations participated. Subsequent meetings were held in 1958 and then at six-year intervals, and the IAEA gradually took over the activities of the Fund.

Meanwhile, E. Blythe Stason, then dean of the Law School of The University of Michigan and managing director of the Fund, had been studying ways of modifying the Atomic Energy Act sufficiently to permit private industry to proceed while still protecting the vital national defense for which the government had sole responsibility. The 1945 act had put the entire responsibility on government agencies and created a governmental industry. Private business could take no initiative but could only act as contractors for the agencies. As a result of the efforts of hundreds of people, the Atomic Energy Act was amended in August, 1954, to permit the AEC to distribute fuel under licenses and to control the safety and technological standards in the construction of power plants. No private ownership of fuel was allowed. It was kept under strict government inventory to prevent it from falling into unauthorized hands. President Eisenhower was at the summer White House in Denver when he signed the new bill into law. I learned about it in Ankara, Turkey, where I was on a mission for the Marshall Plan.

The Fermi Fast Breeder Reactor

The AEC then invited proposals for the building of demonstration plants, and three different plans were offered. Detroit Edison and its associates advocated the liquid sodium fast-breeder reactor to be built near Monroe, Michigan. Senator Clinton Anderson said, "Cisler, when are you going to stop talking about this and tell us you are going to build it?" Roy Searing, president of Consolidated Edison, came down from New York and said, "We'll build one without government money also." That was how Indian Point was established as a pressurized water reactor with Westinghouse as the contractor. Commonwealth Edison presented the third plan. Along with General Electric, it proposed to obtain the private capital needed to build the Dresden boiling water reactor at Morris, Illinois.

Under the amended law it was now possible to proceed with drawings and specifications for a nuclear power reactor. Atomic

Power Development Associates (APDA) was incorporated in New York to do the research on the new technology. Thirty-three companies joined in this effort. Shortly after, Power Reactor Development Company (PRDC) was incorporated in Michigan to build and operate the Fermi plant. Twenty-five companies from APDA joined to put up the capital. The site near Monroe, in the southern end of Detroit Edison's service area, was close to the Toledo and Ohio power companies and near the Consumers Power Company that served much of Michigan. The entire project, including its problems and potentialities, was fully discussed on a number of occasions with the Michigan Public Service Commission. Because of its benefits to the state, the commissioners were in full agreement to proceed.

As president and member of the board of APDA and PRDC, I signed the letter applying to the AEC for a permit to begin construction. This was issued on August 8, 1956, eleven years after the passage by Congress of the Atomic Energy Act. President Eisenhower wrote a personal letter reminding us that the first international conference on the peaceful use of the atom had taken place in Geneva, Switzerland, in August just a year earlier. On that occasion, Lewis Strauss had announced the Fermi project to counter the Russian breeder program.

The AEC issued a provisional certificate for construction, and Strauss came to Detroit for the ground-breaking ceremonies. In his talk he referred to the public-private power controversy in the national government and noted that there was opposition to the free enterprise development of nuclear power in this country. But the preliminary steps had been taken.

Before the war Strauss had been a successful financial expert in New York. As an influential member of the Republican Party and of the first AEC, he became an atomic energy adviser to Eisenhower, who appointed him chairman of the AEC.

Opposed to him was Clinton Anderson, Democratic senator from New Mexico and chairman of the Joint Committee on Atomic Energy. From 1945, the joint committee and the AEC had competed for authority, but with a Democratic Congress and with the first Republican president in twenty years, this was now intensified along political lines. In addition, both Strauss and Anderson were strong-minded individuals with personal reputations at stake. Communication between them was difficult, and at times they would not travel together.

Anderson did not come to the Fermi ground-breaking, so Strauss dominated the news.

Up until 1955, I could not have had a better supporter than Clinton Anderson. But the social and legal problems surrounding atomic energy that took up much of the next five years were almost as difficult to resolve as the technological. Senator Albert Gore, reflecting his long experience with the federally financed and owned TVA, wanted Congress to authorize $400 million to build plants that would then be turned over to private industry. Senator Anderson supported the Gore bill, and I testified against it on the ground that the government already had total control of nuclear fuel and that the private sector needed to be able to put its capital and its expertise into the commercial development. Had the bill become law, the small body of experienced nuclear scientists and engineers would have been lured into the federal program to join the others who were working on military applications.

Nevertheless, in the proceedings Anderson was alienated and I fell into disfavor. As a powerful Democrat he got to some of the political leaders in Michigan. Later he told me that he was not really opposed to me or the Fermi plant but to Lewis Strauss. "When I start swinging an ax," he said, "I never stop until it lands. You just happened to be in the way."

My experience up to this time had reinforced my early belief that the American people gain the greatest benefits from a mixed economic system with a balance between government and private business enterprise. This was analogous to the checks and balances we have built into our political system. Although this position created some opposition in Washington, it also pulled strong forces together in the free enterprise sector of the economy, and the development of nuclear energy did move ahead in the nation.

There were further legal difficulties to be resolved in Michigan. Questions about the safety of the Fermi plant were raised, and Governor G. Mennen Williams named a committee on atomic energy under the Public Service Commission. After an investigation, this official body reported that it was satisfied with projected safety measures. Then three labor unions intervened with the AEC and asked that the construction be stopped. Among many expert witnesses both Dr. Hans A. Bethe of Cornell University, who had won a Nobel prize in nuclear physics, and Dr. Norman Hilberry, Director of the Argonne National Laboratory, testified for the project. This government agency was developing a 16,500 kw Experimental Breeder Reactor (EBR-2), so it had experienced people. After three years, the commission ruled in favor of Fermi, noting that the plant was subject to final safety checks when construction was completed.

The unions then took this decision to a three-judge Court of Appeals, which ruled two to one that all questions about safety should be satisfied first. The practical effect of the judgment was to create a situation in which the plant could not be built until it had been proved safe, and it could not be proved safe until it had been built. So the decision was appealed to the full nine-man Court of Appeals, and the AEC joined the action, arguing that the court had substituted its own opinion on the highly technical matter of nuclear safety for that of the legally constituted authority. The Court of Appeals, however, upheld the earlier decision.

Since the authority of the Atomic Energy Commission to regulate had now been called into question, the case had taken on constitutional importance, and the solicitor general of the United States entered it, along with the AEC and the Power Reactor Development Company. Finally, when the United States Supreme Court ruled in favor of the AEC on June 12, 1961, the last major legal problem had been resolved. It was almost five years after the commission had issued the permit to build.

I later became a friend of Walter Reuther and worked closely with him in many civic projects in Detroit, but I never really understood why the UAW took the leadership in opposing clean and inexpensive energy. It seemed clear to me that any unnecessary manufacturing expense merely added to the cost of material and so was not available for wages and salaries. A 1,000 mgw nuclear plant such as the company began to build in the late 1960s uses one ton of uranium a year and leaves one ton of waste. A 1,000 mgw fossil plant uses three million tons of coal that must be hauled hundreds of miles and leaves 30,000 tons of ashes and waste. Since Michigan was an energy-importing state, this posed a particular hardship on the very companies which were providing jobs and economic benefits to members of the union. Perhaps the UAW opposition arose because the technology was new, was not fully understood and hence feared even by the industrial unions, and because atomic energy had been associated with the bomb.

Fermi #1 was designed to produce 67,000 kw of electricity

with its first core, one-fifth as much as the conventional coal-fired plants of the early 1950s. Still it was large enough to constitute a demonstration rather than an experimental model, and until it was phased out in 1973, it was the largest breeder reactor in the world. Since it was built to test the economy of nuclear generation and the feasibility of the plutonium cycle, the original plan was to sell any electricity that was produced to The Detroit Edison Company. The revenue from this would return to the Power Reactor Development Company which owned the reactor. Under the terms of the original contract, the Atomic Energy Commission agreed to provide $4,450,000 to underwrite some of the research. It also helped train the operators, and it leased the uranium fuel for five years without charge. The remainder of the financing came from the member companies in the consortium. This seemed to be a workable balance between public and private interests.

The reactor developed several mechanical difficulties during the testing period before the nuclear fuel was added. The graphite shielding had to be replaced, the fuel assembly system had to be redesigned, and the liquid sodium system had to be modified to eliminate a check valve closure problem. These steps were time-consuming and expensive, but the problems were eventually solved, and a wealth of important information was gathered. Finally, after more public hearings, the AEC on May 10, 1963, approved the safety requirements, and the fuel elements were loaded into the reactor. At 12:23 P.M. on August 23, the reactor began to operate at the one thermal megawatt power level that was authorized. Over the next several years and after extensive testing, the rate was increased in regular stages and met the inspection standards of the AEC. In August, 1966, Fermi was operated continuously over a fifty-five hour period at half its rated power level and produced one million kilowatt hours of electricity.

Then followed the only incident that involved the nuclear fuel. A small flow guide in the bottom of the reactor vessel broke loose and was carried by the liquid sodium until it became fixed in a position to obstruct the flow. This caused a partial meltdown in about one percent of the core and the emergency safety procedures shut the plant down. The greatest care had been exercised in the engineering and construction so the safety systems worked as they had been designed to do. The structure alone had been built to contain the possible melting of the entire core. It took three years to correct the mechanical failure because the reactor was now radioactive and required special handling. Even this incident, costly as it was in both time and money, demonstrated several important factors about the fast breeder, perhaps the most important being the ability to resume operation after a disruption of some of the reactor fuel. Fermi was again authorized to operate in 1969, and did so for thirty full-power days using the second core that was available at the site.

Nuclear experts and the general public share a deep concern about the safety of atomic energy. The scientists and operators who work with it closely are naturally the most aware of the hazards. A nuclear power plant cannot behave like an atomic bomb because it is not built like a bomb. Like the automobile and any other technological development, however, the plants are potentially hazardous and must be run carefully. Twenty years earlier I had been involved in superimposing a mercury cycle on the normal steam cycle at the Kearny power plant in New Jersey to improve the fuel efficiency. As a metal, mercury has good heat transfer qualities, but it is also highly poisonous and must not escape into the environment. So I was very aware of the need for safety, and we fortified all the Fermi systems and built one containment barrier around another. When the partial meltdown did occur, the safety elements acted in the way they were supposed to and shut down the reactor.

As a result of careful design the power plants have an outstanding safety record. With more than fifteen years of operating experience there has not yet been a claim for bodily injury or property damage caused by the fissionable material in either commercial power plants or demonstration reactors like Fermi #1. The federal government has an insurance program paid for by the utilities to protect people, and so far it has not paid a single claim. Even related concerns for the health hazards of possible leakage of radioactive material have abated as the excellent safety record continues year after year.

In addition to the tremendous amount of knowledge about this new source of energy that Fermi #1 has produced, it also attracted and provided an education for a large number of able men, a number of whom became corporate officers in Detroit Edison. Robert W. Hartwell, who was the first project engineer at Fermi and who had worked with the Rickover group on the nuclear submarine program, became senior executive vice president for finance. Walter J. McCarthy, who had studied at Cornell and had joined Public Service of New Jersey as I had done many years earlier, joined the Fermi project. Hans Bethe was investigating the physical properties of a meltdown and whether, even theoretically, the elements could be reassembled to produce an explosion. It was learned that they could not. Along with this senior scientist, McCarthy was responsible for the safety program of the project. He became executive vice president in charge of Detroit Edison's engineering, construction and power systems operations. The project made them, and many others, think hard and logically about an entirely new range of problems. It made them discipline themselves to undertake difficult decisions and to gain confidence as they solved a myriad of concerns—scientific, technological, financial and social—that no men had faced before.

The Fermi plant attracted people from all over the world. In addition to hundreds from the United States, the British sent engineers for extended periods of study as did the Belgians and an international consortium of European countries from Euratom. The British have built a 250,000 kw liquid-metal fast breeder, and the French have completed the most recent one which is operating very efficiently. The Japanese sent many men to Detroit to build on the Fermi experience, and a number of them spent several years here with their wives and children. Today all the useful parts of the plant have been shipped to Japan. Engineers from the U.S.S.R. also studied the plant and have now constructed a 350,000 kw fast breeder, the largest to date. In 1973, when Fermi #1 was decommissioned due to lack of funds for a new core, the United States was the only major industrial power in the world without such a reactor in operation or in the advanced design stage. It will not have one now until the 1980s. This is a major disappointment to me.

I am still convinced that the breeder should have been given a higher national priority, but in the end the Fermi group was the only one to put money into it while the rest of the private financing went into the slow neutron technology. The water-moderated reactors were easier to design and the AEC decided to concentrate development money on them. This saved the operating companies the R & D costs that had to be paid out of earnings and enabled them to build the large commercial plants that were financed as long-term debt through the sale of stocks and bonds. In the short run it was desirable to go commercial as quickly as possible.

In the long run, however, they will be expensive. They are at best a stopgap model that will run out of fuel by the end of the century. And they are inefficient. A pound of uranium, intrinsically equivalent in energy to about 1300 tons of coal, can be utilized only to the extent of about 2 percent or twenty-six tons of coal, in water reactors. The breeder can increase the utilization thirty times to sixty percent. Because of this efficiency the projected use of the breeder over the next half-century could save as much as three billion tons of coal and $200 billion.

The demands of the people of the world for more electricity have not abated, and they have not waited for the maturation of the fast breeder reactor. Before the end of the 1960s, three commercial, water-moderated reactors were operating in the United States, and soon after this Detroit Edison also began to build one that would be almost twenty times larger than Fermi #1. Environmental concerns have forced serious delays. Whereas, in most countries a nuclear power plant can be brought on the line in four or five years, in the United States it now takes ten years to do the same thing because of governmental regulations. The cost of tying up hundreds of millions of dollars in non-productive use over ten years is tremendous and will be paid for in much higher rates for electricity than would have been necessary had we followed the example of other nations. This will give them a significant trade advantage which will again be paid for by the American consumer.

But Fermi #2 will come on the line because it must. In 1974, the year of the World Energy Conference in Detroit, seventeen member countries already were using uranium for power generation and ten others planned to have operating plants by 1980. The United Kingdom could generate 6,000 mgw using twenty-nine reactors. The United States was far ahead of the rest of the world with fifty-five operating reactors capable of producing 37,000 mgw and with sixty-three more under construction.

When the push to develop a new source of energy started thirty years ago in 1945, many businessmen like me thought of it as a supplement to the large coal resources within the continental limits of the United States. No one foresaw the extent of the environmental objections to coal because power companies like Detroit Edison had been bringing air quality controls into their systems since the 1920s. And no one anticipated that the cost of coal would triple over a three-year period.

No one foresaw the embargo arising from the Arab-Israeli conflict that was starting in Palestine or the sudden inflationary increase that quadrupled the price of oil twenty-nine years later. A gradual price increase was anticipated because the known reserves of oil were quite limited in relation to the rising demand. As it became more difficult to find new fields and as the costs of drilling wells in the sea beds and transporting it over long distances increased, the cost of nuclear energy would become more attractive. In 1975, atomic energy suddenly became as essential on the civilian front as it had been on the military. The sharp increase of over one-third in the rate of construction around the world was largely the result of political uncertainties arising out of the oil embargo and the economic certainties arising out of the tripling of the price of fossil fuels that made uranium a most attractive competitor.

The scientific and engineering problems of the conventional nuclear power plants are now largely resolved. Some work remains to be done to bring the breeder beyond the Fermi demonstration stage to full commercial size but, in general, it can be said that the breeder is ready for the United States any time the country wants it. Looking back at the *Synchroscope* article of 1941 on "U-235 . . . What is it?" I discovered a notation on each of the two pages. One stated: "Ignorance is the mother of fear." The second stated: "The poverty of earth is never dead." We have come a great distance into the atomic age since 1941, but those two statements still haunt the world.

Colonel Cisler, Chief of the Public Utilities Section, Supreme Headquarters Allied Expeditionary Force, European Theater of Operations, 1945.

Colonel Cisler, left, and Julius Krug, head of Office of War Utilities,
after World War II.

The Cislers at home, 1953.

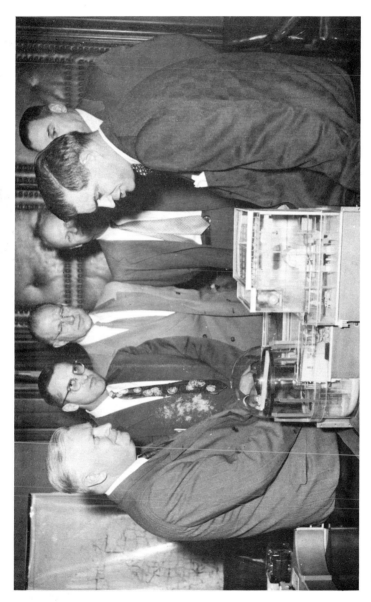

Cisler, left, and Governor G. Mennen Williams of Michigan, right, with a model of the Fermi fast breeder reactor, 1956.

Cisler presents Mrs. John Eyre Sloane, daughter of Thomas Edison, to Crown Prince
Akihito and Crown Princess Michiko at the international Edison birthday celebration
in Tokyo, 1964.

Indian Prime Minister Jawaharlal Nehru and Cisler, 1961.

Left to right, Shah Mohammed Reza Pahlevi of Iran, President Lyndon B. Johnson, Walker L. Cisler, Mrs. Cisler, September 17, 1965.

Cisler shows off a brace of ponies at his Gradyville, Pennsylvania farm.

An easy rider, 1962. Cisler habitually cycles five miles before breakfast.

Cisler greets shareholders at Detroit Edison's annual meeting, 1968.

Cisler pauses for a drink at an outing of the Power Club, a Detroit Edison employee organization, 1974.

Cisler and his foreign medals.

CHAPTER VI

INTERNATIONAL AFFAIRS

The Marshall Plan

On June 5, 1947, three years less a day after D-Day, General George C. Marshall gave his famous talk at Harvard University about the reconstruction of Europe and the need to bring together a variety of assistance programs that had developed since the end of the war. He had retired as chief of staff of the armed services and was serving as secretary of state to President Harry S. Truman. This was the beginning of the Marshall Plan which, in a variety of forms and under many names, was to set the direction of relations between the United States and Europe that is still going forward. The speech stated that the United States would cooperate in helping the Europeans help themselves through a program designed by Europeans for a number of European nations. Our policy would not be directed against any country but against hunger, poverty and desperation. The aim of the plan was to revive a working economy in order to encourage the emergence of political and social conditions in which free institutions could exist.

Ten months later, after much deliberation and under the able leadership of Senator Arthur H. Vandenberg of Michigan, Congress adopted the Foreign Assistance Act to underwrite the European Recovery Program for a period of four years at an estimated cost of $5 billion. Western Europe had already recovered sufficiently its industrial output to begin to approach the prewar level. But standards of living were still low, and there was a wide disparity from country to country. Paul G. Hoffman was named head of the Economic Cooperation Administration with virtual cabinet rank. He, along with W. Averell Harriman, who had charge of the program in Paris, invited a number of businessmen to help implement the decision of the Congress.

Meanwhile, a group of European leaders had begun to plan along the lines suggested by Marshall and on April 16, 1948, the Organization for European Economic Cooperation (OEEC) was established. The sixteen member nations agreed to set production targets, improve the use of plants and manpower, move toward fiscal and monetary stability, and discuss customs unions. The American position favored private enterprise, and the European participants agreed even though the Communist pressure for state control was particularly strong in some of the countries.

The Cold War had started. The U.S.S.R. had announced it would keep its grip on eastern Europe, and Winston Churchill had described the Russian position as an iron curtain in his speech at Fulton, Missouri. The Greek Civil War had broken out and the Truman Doctrine, promising aid to resist armed aggression in Greece and Turkey, had been proclaimed. Later, in 1947, the Communist Information Bureau (Cominform) was established in Belgrade to coordinate activities, and this was followed by the coup in Prague when Czechoslovakia was added to the satellite nations. In June, 1948, the blockade of Berlin was begun and the eleven-month air-lift into the city was started by France, Great Britain and the United States. These political activities produced ideological consequences. French Communist Party leaders pulled maintenance men out of the coal region and caused the mines to flood. Not even during the war period, in the retreat of the Germans, was such violent action taken. It caused months of delay in the production of coal that was vitally needed for economic recovery.

Jim Parker was very much concerned about resuming construction at Detroit Edison to make up for the loss of time during the war years, but when Hoffman phoned, he allowed me to go to Washington. I explained my wish to proceed with the job I had been hired to do in 1943 and had delayed so long because of military service. Hoffman described what had happened to the electric power situation since I had left the Allied Military Government at the end of 1945, and the need to organize it if the economic life of Europe was to progress on the long road to recovery. I knew how inadequate the utilities were, but I also knew how serious were the needs of my own company. Finally, Hoffman queried the power people in Europe who said that if Cisler would undertake this assignment they would give full international cooperation. Jim Parker agreed that the obligation of the United States had to be met so I took on this additional task.

Although I had to work night and day so as not to neglect my responsibilities to Detroit Edison, I found my assignment with the Economic Cooperation Administration very satisfying. We sent people from American companies to Europe to survey the situation and to help. We brought European engineers to the United States for study and training here. During the time of allied military government we had established transmission interconnections. I worked with people at the local level as often as possible. I made it a point to be present, for example, when they closed the switch to join Belgium and Germany together. The Union for the Coordination of the Production and Transmission of Electric Energy (UCPTE) was formed, and now under the Marshall Plan this was extended. I wrote two memoranda: one for the interconnection of the northern countries and the other for the southern, and for the coordination between them. Although this step had never before been taken in Europe, it functioned because people had seen what could be accomplished.

Switzerland had hydro power that could be sent to France and Germany. Belgium and the Netherlands had coal generators. The Germans had private companies and the French had nationalized their system. But everyone worked together for the common good. Questions of political nationalism were simply avoided. The engineers re-established the systems, and electric energy flowed back and forth across boundaries on an hourly basis without import or export licenses. All that was required was an accounting system so that the organization that incurred the expense of generation was recompensed for it.

The Italians under Mussolini had relied almost completely on hydro power, and as a result of our studies they started to add coal-burning plants, and they trained a number of their engineers at our plants in Detroit. It was at this time that I met Arnaldo M. Angelini, who had been a professor of electrical engineering at the University of Rome and then had become general manager of one of the private companies. Many Italian companies, however, failed to keep in touch with their customers and within a few years Italy's power systems were nationalized. I worked with Angelini not only on the Italian system but also on the power supplies and interconnections in

Western Europe. Today he is chairman of Ente Naxionale per l'Energia Ellettrica (ENEL), the largest operating company in the country.

In Greece, only Athens and Salonika had power capacity at all commensurate with their needs. We found additional dam sites and some lignite supplies that could be developed. With the aid of American dollars and experts, along with the skill and manpower of Greece, these resources were improved. More than 1,200 miles of power lines were constructed. The entire program cost about $100 million, financed by United States dollars, counterpart funds and money from the reviving Greek economy. It was at this time that I first met Constantinos Doxiadis, who was planning the Greek reconstruction, and over the years my respect for his ability grew steadily. He built more housing than any man in history up to that time, and he survived as a member of the government through change after change of political leadership.

Altogether, I visited eight countries, seven of which I had known during the war, and I had the pleasure of meeting many of my former Cisler Circus Associates. There were 43,000 mgw of capacity in operation and an additional 1,000 mgw were placed in service in 1947. In order to double the capability in a seven-year period, however, additional help was needed from the United States, and I was there to assist in this effort. I even flew into Berlin one night on an air-lift for ten tons of flour.

Building power plants was immensely more satisfying to me than seeing them destroyed, and the design and creation of international utility systems broadened my understanding of the importance of electricity. It did much of the work that fed people, clothed and housed them, and made their lives more comfortable than in my youth. It also permitted them to operate in many ways without the old constraints of national boundaries and jealousies. The language of the industry was not English, French or German but was universal, and a kilowatt hour was understood to mean exactly the same thing in Turkey, at one end of the Organization of European Economic Cooperation, as in Iceland at the other. Everywhere the people involved in this activity worked hard to make things better rather than to gain advantage of one another. If electricity could be made more cheaply in one country, the benefits would be passed on to its own people and then be allowed to flow across the interconnections to help others.

Many of the established leaders tried to advance the principles of political unity at this time, and in the spring of 1949 the Council of Europe was established. Winston Churchill had called for a United States of Europe as an act of faith in the European family, but there were many national differences. Two years later the Belgian statesman Paul Henri Spaak resigned as president of the Consultative Assembly. But in April, 1949, a more limited military agreement was reached when Canada, Norway and the United States joined the Brussels Treaty countries of Western Europe in the North Atlantic Treaty Organization (NATO). Five months later the U.S.S.R. announced that it had built an atomic bomb, and the balance of terror had begun.

All my activities in international matters were conducted in the pervasive atmosphere of the Cold War. Its existence was as much taken for granted during the period as the existence of air is assumed in the process of breathing. Thus we found ourselves in a world torn with conflicting ideologies, desperately in need of inspiring leadership.

The foreign policy of the United States was directed toward containing the forces of the U.S.S.R. and other Communist nations. This led us to establish military bases around the world and to continue to improve the atom bomb and later the hydrogen bomb. Such weapons and their delivery systems massed the greatest quantity of destructive energy ever put together.

So the suspicions and hostilities that had flamed out in World War II continued as a fact of life that had to be considered in any reconstruction plan. I had little contact with the politics of international relations, nor did I have much to do with political theories or ideologies. But I did have a great deal of association with government, with the management of public institutions, and my experience with a regulated industry had prepared me to cope with this. All over the world the business of government is conducted in bureaus, the people who manage them are service-oriented, and many of them are both well educated and dedicated to the public good. In general, I have worked with that part of the private sector that is production-oriented. What I attempted to bring to the international as well as to the national enterprise was improved productivity so that more electric service was delivered. Both public and private institutions exist to bring a variety of goods and choices to individual people. Because of this background and these beliefs, I concerned myself less with political and more with economic matters.

Energy Studies in Europe

In the 1950s I began to understand the working of economic systems better than I had before. From boyhood I had been immersed in the growing industrialization and urbanization of the United States, but I took part in the process without being really aware of more than pieces of the grand design. I lived in cities while the countryside of my youth was gradually being absorbed into the daily urban system of Philadelphia. I chose to go into a profession instead of becoming a farmer, and I selected engineering with its emphasis on quantitative thinking. This led me into industry and, through a combination of factors, into the power industry. As industrialization again moved ahead after the war, I found myself dealing with the most sophisticated form of energy for an increasingly complex, industrial way of life. Europe, of course, was the home of the industrial revolution and was a good place to observe the process of bringing it under control.

When the movement toward a federated Europe faltered, people began to develop organizations that would encourage economic cooperation. The leadership of OEEC and others created the European Coal and Steel Community (ECSC) with a Common Assembly, a Council of Ministers, a High Authority chaired by Jean Monnet, and a Court of Justice. Overcoming such problems as what official language to use and in what country to locate the headquarters, ECSC set out to lower trade barriers, expand the economy and raise living standards, and move toward the creation of a Common Market. It was a supranational effort and it met with growing success.

For me, there was an immense span between the Alexander Hamilton Institute textbooks of 1925, the efficiency studies of machinery and power plant operations of 1935, and the Harold Hartley study of 1955-56. By the middle fifties, the Organization for European Economic Cooperation included seventeen nations which were concerned about the failure of growth of the coal industry to match the increasing production of iron and steel. OEEC appointed an eight-man commission from the member nations with Sir Harold Hartley as chairman, and I met with them. Over a period of ten months, the staff of the commission examined the use of energy since the 1920s, interviewed representatives of energy producers, trade-unions and employer organizations, and each member obtained from his own country the best projection of probable energy production and consumption for the next twenty years.

The Hartley report was published under the title *Europe's Growing Needs of Energy—How Can They Be Met?* It emphasized the importance of coal as the mainstay of the energy economy for many years to come and noted that alternative primary fuels like oil, gas and uranium could be developed to fill the gap between supply and demand. There was talk of importing 50 million tons of coal a year from the United States but the economies of oil were more favorable. As president of the World Power Conference, Sir Harold was in a position to make people listen to the conclusions.

Shortly after the publication of this report, another team was pulled together under the leadership of Austin Robinson, a professor at Cambridge University and Secretary of the Royal Economic Society. Once again I was the United States observer and found myself among old friends like Arnaldo Angelini and Herman Abs of Germany. We discovered the OEEC countries rapidly converting from coal to the less expensive oil from the Middle East. This depressed the coal markets of Europe, creating unemployment and political unrest. Then, when the Arab-Israeli conflict broke out between November, 1956, and June, 1957, practically all oil movement stopped. United States marines landed in Lebanon to try to stabilize the situation around the pipe line, and when Sir Anthony Eden ordered the British fleet to go to the Suez Canal, Eisenhower ordered the American fleet in the Mediterranean to interpose itself. The political crisis passed, but the fears it created and the recession of 1957-58 spurred the search for nuclear energy. A crash program began in England to develop the gas-cooled reactor.

On the Continent, intensive investigations into the civilian use of atomic energy had begun in 1952 with the founding of the European Nuclear Research Council. Its activities paralleled many of our own. Two years later the International Atomic Energy Agency was established, and in 1955 I chaired the first meeting of the Conference for the Peaceful Uses of the Atom in Geneva, Switzerland. The Suez crisis, however, increased the sense of urgency. The men associated with the OEEC created another supranational organization, Euratom, to coordinate the research and development of nuclear power plants and named a five-man commission to run it and a 142-member

Common Assembly to which the commission reported. The European Community was spending $2 billion a year to import energy supplies with predictions that this would double in a decade. Oil was still the cheapest form of available energy, however, and exploration was pushed in Venezuela, Libya and the Sahara, all outside the trouble area. Additional pipelines were built and very large tankers carried oil inexpensively from the Persian Gulf all the way around Africa, and gradually the push for nuclear generation slowed again.

The confidence gained in international economic cooperation through OEEC, the European Coal and Steel Community and Euratom enabled the "Inner Six" nations of Belgium, France, Germany, Italy, Luxembourg and the Netherlands to establish the Common Market in 1958. Although the market basically broadened the customs union of the coal and steel authority to include a full range of products, it quickly moved to many other fiscal and monetary matters to encourage free trading within the Community. Once again, its officials represented supranational interests.

The Robinson report was published in 1960 under the title *Towards a New Energy Pattern in Europe.* Perhaps the most significant finding grew out of the forecast that between 1955 and 1975 the investment for energy requirements would be $130 billion and for all new capital would be $1,100 billion. Subsequent studies confirmed this ratio in all countries, whether mature industrial nations or developing ones. The investment required for the production and distribution of coal, electricity, fuel oil, gasoline, natural gas and other energy commodities accounts for 12 percent, or one-eighth, of the total investment required for all purposes. The other seven-eighths were required for farms, homes, industrial and commercial plants and transportation systems. As a boy I understood about wages of a dollar a day and stoking a furnace and emptying the ashes by hand. Because of the energy available to substitute for hand-power, life was a great deal more varied, complex and comfortable sixty years later.

Energy Studies in the Pacific Orient
While the Marshall Plan directed its attention to Europe, President Truman also started the Point Four Program for other parts of the world in need of help. Under General Douglas MacArthur, the Japanese people quickly began to put their lives in order again. Public health services were restored

immediately after the war and the transportation system was rebuilt and extended. The major wealthy families which had emerged out of feudal Japan into modern history were restricted, and both land and industrial wealth were distributed more widely than before. The electric power industry, however, was composed of many small companies with small generating stations and transmission lines serving local geographical areas. This was inadequate for the needs of the times.

One day in January, 1949, Robert Lovett, who was secretary of the army, phoned to talk about the situation, and I joined him in Washington to write the orders to send Joseph M. Dodge, president of the Detroit Bank and Trust Company, to Japan. We had to word these orders very carefully so General MacArthur would not be offended or feel that he was being coerced. I felt that he would respond favorably because I had helped him earlier by sending one of the floating power plants to the Philippines after he had returned there. A week later a message came back requesting Dodge.

I had known Dodge first in Berlin in 1945 when he was financial adviser to General Lucius Clay and began the reorganization of the banking system and the reform of the currency through the exchange of ten old reichsmarks for one new deutsche mark. Once again he had to tackle the problem of inflation. He persuaded Japanese businessmen and government officials to introduce more planning, stabilize the yen, reduce taxes and start building a broader industrial base. When he returned to Detroit in May, he had prepared the first balanced budget since the invasion of Manchuria, a generation earlier.

After the austerity of the war years the people really needed consumer goods, but they disciplined themselves in traditional Japanese fashion and postponed satisfying their needs for a while. Doug Campbell and Bob Van Duzer of Detroit Edison participated in the first electric power survey, which was published with the New York City address of the Edison Electric Institute on it, and the Japanese electric utility industry was organized into nine large geographical areas. Tokyo Electric Power Company absorbed thirty-three smaller companies and is now the largest investor-owned public utility in the world.

The method of making an electric power survey of the kind we had developed at this time for the Edison Electric Institute was systematic; but compared to the comprehensive and total energy study that was produced ten years later in the Robinson

Report, it was relatively simple. It began with an inventory of existing plants and lines, and then it moved to a projection of future growth of population, industries and markets. We surveyed the native energy resources and worked out the best balance among fossil fuels and water power. Existing charges for electricity were studied in the context of a plan for both area and industrial development. We looked at the capital requirements and how fuel imports would influence the balance of payments of the country. We also considered the organization of the local power industry and the technical and managerial skills that would be needed to run the new system.

Studies of this dimension were particularly valuable in helping to revive the industrial nations of Europe and Asia that had been devastated by the war. They provided guidelines for the longer reconstruction process. Electric energy had become such a critical element in advanced industrialization that it had to be provided before any substantial progress could be made.

Japan recovered very quickly and The Detroit Edison Company gained directly from its contacts on many occasions. The first came when we were extending the capacity of the Trenton Channel Plant and discovered that high pressure pipe was scarce. We finally got a bid for $700,000 but with an indefinite delivery date far into the future. Chester Ogden, who would soon be named manager and later vice president of purchasing, filled our order in Japan for $293,000 and a firm delivery date of six months. Almost twenty-five years later those pipes are still performing satisfactorily.

When the Korean War started in July, 1950, our European friends in OEEC were concerned. NATO had only 14 divisions and 1,000 aircraft while the USSR had 175 divisions and 20,000 aircraft. France was deeply committed in Indochina. The outbreak of fighting in the Pacific created further uncertainty in the Cold War.

The Korean War gave a considerable boost to the Japanese economy, but it was destructive to Korea itself. A basically agricultural country, its limited resources were divided between north and south. The Yalu River hydro electric dam and plant had been built by the Japanese, but as American forces pulled back south of the defense lines on the 38th parallel, this major source of electricity for the south was cut off. I got a call from the United States Agency for International Development (USAID) office in Washington and quickly helped put a team together for our government as I had first done in 1948. At that earlier time, the facilities built by the Japanese were fundamentally adequate to the needs of the country, but they were in poor repair. The Pusan generating plant was supported by the floating power plant *Jacona* with 20,000 kw of generation. The floating power plant *Electra* with 6,000 kw was anchored in Inchon. Edward Ralls of Detroit Edison was the first of the team to go into the plants. By providing repairs and some additional diesel-generating sets, the system was gradually restored.

During the Korean War the Japanese, with their engineering skill and knowledge of the language, provided supervisory personnel. The old antagonisms between Japan and Korea were minimized. We assembled kits of wrenches in Detroit because they did not have tools to work with. In the field, the men used shoelaces to pack steam valves. Another of the old floating power plants was moved to Pusan to supply that city and the United States military forces with electricity. We trained 132 Korean engineers on Edison equipment in Detroit. Once again, through cooperation and mutual concern, we were able to provide electricity adaquate for military needs and for laying the basis of sound industrial growth.

The work in Korea continued for many years, and from time to time Detroit Edison men were sent there under the utilities division of USAID. Howard Canfield visited the country off and on for a quarter of a century. By April, 1964, the Korea Electric Company announced an end to all restrictions on the use of electricity and advertised in the Seoul newspaper that people could now have appliances in their homes as well as electric lights.

I did not go to Japan myself until 1956—for a discussion of nuclear reactors. The energy studies going on in Europe were reinforcing my conviction that we had to find substitutes for fossil fuels. The State Department asked me to brief the American ambassadors in Paris, including our first woman ambassador Clare Booth Luce, on atomic power. I had a small model of the Fermi plant prepared and added to my energy box a bar of uranium weighing 3¼ pounds. I explained that the slow neutron reactors that were moderated by water or gas could use from 1 to 3 percent of the energy in natural uranium. With the use of the fast neutron reactor like Fermi, which bred plutonium at the same time it used the fissionable uranium, we could recover up to 80 percent. The bar of uranium in my energy box could generate 12 million kwh of electricity. A piece of coal of the same weight could generate 4¼ kwh even with the most advanced power plants that had been designed.

After the wedding of our daughter Jane to Albert Eckhardt in Paris, my wife and I flew to Bangkok where she went first to Hong Kong and on to Japan. Meanwhile, I went to Tokyo via Manila because I needed to talk with government officials. John Foster Dulles was developing the South East Asia Treaty Organization following the French loss of Indochina, and the Philippines were an important element in the new American-Pacific defense system. In Japan, several of the men who had worked on the power survey met me at Haneda airport at 1:30 A.M. They arranged for me to meet with Matsutaro Shoriki for two hours the next afternoon to discuss nuclear plants. He was head of one of the largest newspapers, the *Yomiuri Shimbun,* and of Japan's Atomic Energy Commission, but he had never seen anything like the Fermi model. The British had developed the Calder Hall gas-cooled reactor and Hinton, as head of British Atomic Energy Authority, was trying to sell it in Japan.

The attitude about nuclear energy was quite negative because Japan had been the only country in the world to have been under atomic attack. In addition, atmospheric testing of so-called dirty bombs, those with a high radioactive fallout, had recently been conducted at Eniwetok in the Marshall Islands. The Japanese were accustomed to fishing those waters and one of their boats, the *Lucky Dragon,* had been contaminated. This incident produced widespread revulsion in the country. But I spoke in Tokyo and Osaka on the work of the Atomic Industrial Forum and of the Fund for Peaceful Atomic Development, both dedicated to using nuclear products for constructive work, and my energy box was as helpful to them as to the American ambassadors. Since they had to ship much of their coal all the way across the Pacific Ocean, the significance of the density of energy in the 3¼ pounds of uranium was clear.

The following year I returned to help establish the Japanese counterparts of the forum and the fund. The United States government could not talk convincingly because its record of atomic destruction was too great, but private businessmen could work cooperatively. By the middle 1950s the Japanese power industry had become one of the most modern and efficient in the world as it supported the extraordinary economic recovery of the country. Ryotaro Takai, an engineer who had trained for a short time before the war at Georgia Power, became president of Tokyo Electric Power Company. He had given a report at a meeting of the Edison Electric Institute in 1951 when the AEC announced that the small Experimental Breeder Reactor (EBR-1) had confirmed the principle of breeding by actual test, so he was in at the beginning of nuclear power. Kazutaka Kikawada, who later became chairman of Tokyo Electric, also participated in the formation of the Japanese industry and in the careful approach to nuclear generation. He had studied economics at Tokyo University and brought a large body of financial knowledge to the industry. Yoshishige Ashihara helped organize the second largest company, Kansai Electric in Osaka, and became its chairman. And the grand old man of Japanese electric power, Yasuzaemon Matsunaga, who worked originally with Joe Dodge to formulate the national plan and then along with others to develop the nine companies under General MacArthur, became chairman for many years of the Central Research Institute of the Electric Power Industry.

The organizations encouraged Japanese engineers and economists to study in the United States, and at the Fermi plant alone we had a series of training contracts worth several million dollars. Some of the men stayed in Detroit up to five years with their wives and children, participating in American life as well as in the nuclear sciences. Some of them earned additional degrees. After Kozo Odajima died in the plane crash at Hot Springs in May, 1971, Wayne State University conferred the MBA he had just earned, the first posthumous degree the university had awarded. It was a signal gesture of friendship and respect.

High school teachers and students also were supported through these organizations. All of this was done with private money by people who wanted to share knowledge and technology so that everyone could enjoy the fruits of well organized work. The model of Fermi #1 was left in Japan to demonstrate the constructive use of atomic energy. In 1961, I received the Order of the Rising Sun for helping develop the peaceful uses of the atom, and the Japanese continued to advance their nuclear technology. Like all other countries, they relied on Middle Eastern oil to see them through the 1960s. By the 1980s, however, they should have enough nuclear generation to free themselves from much of the tonnage of coal and oil they have had to import.

Hui Huang, a graduate of the Cornell engineering school and then head of Taiwan Power, had visited Detroit to ask my help. We saw to it that he and his executive vice president, Yun Suan Sun, attended the Atomic Industrial Forum meeting in Tokyo. His wife, Lily Sun, and Gertrude found they had much in common. Since I simply could not fit a visit to Taiwan into my schedule on that occasion, I asked Harvey Wagner to go. The USAID Mission wanted to develop the hydro power first because the water was available and it was inexpensive to convert into electricity. Later, I did visit the country when the off-shore islands of Quemoy and Matsu were under bombardment. Every other day the Taiwanese would return the fire and on the odd days would launch propaganda balloons. Secretary of State John Foster Dulles was very concerned, although in time the shelling simply died away. This is when I discussed the energy situation with Generalissimo Chiang Kai Shek and we planned a total study for the island.

Taiwan had suffered heavily in World War II and when the Japanese were forced to withdraw, their engineers and management people went along. In the capital, Taipei, 80 percent of the buildings were destroyed or damaged and the generating capacity had been cut to one-tenth. Capable people with technical skills had come from the mainland, however, and the island had started to recover. The United States advanced an adequate line of credit and Taiwan rapidly became a model for the rapid transition from an agricultural to an industrial economy. In Detroit we trained more than one hundred of these people.

An interesting incident occurred ten years later in 1967 in Nigeria. Y.S. Sun had gone there for the World Bank to help finance a power dam. I was involved with the Nigerian committee of the World Energy Conference and we observed the dam from a small plane. Fighting had broken out between the Biafrans and the Nigerians, and the troops which guarded the dam had not been notified of our inspection flight. They opened up with rifles, and I experienced my first attack in a plane.

When the project was completed and the capital, which was falling apart from lack of power, began to come alive again, Sun returned to Taiwan as minister of Economic Affairs. I mention this to show that there were people all over the world who had confidence in one another and who had a record of being responsive when a request for help came. Although I am here recording my own perception of the international activities of three decades, I was just one among many. They were creative, thoughtful men, with records of accomplishments and dedicated to the public good. Each of them contributed to my store of knowledge and to the satisfaction of working on mutual concerns.

Energy Studies After 1960

In 1958, I went to the Soviet Union for the first time with an Edison Electric Institute team to exchange technical information with our counterparts in European Russia and to visit some of the power plants. Harvey E. Bumgarder of the company was chairman of the E.E.I.'s Committee on Technical Exchanges for Overseas Visitors and was a member of the group. Minister Anastas I. Mikoyan invited us to the Kremlin and talked about our visit. Although he preferred to speak through an interpreter, he understood English, and we had a thorough discussion of electric systems. He realized the United States was way ahead but said, "We will catch up with you."

"No," I said. "We will keep ahead because of our tech—nology and our engineers."

At that time the U.S.S.R. had a total capacity of 53 million kw and the average home used 400 kwh a year. The U.S.A. had a capacity of 167 million kwh and the average home used 3,550 kwh a year. These figures marked the difference between the standards of living of the two countries. We were impressed, however, by the quality of their equipment and their pride in their work. They had come a long way since Hitler had invaded their country and caused such terrible destruction.

The following year Mikoyan visited the Russian trade exhibit in New York and then came to Detroit. There were some local fears and expressions of hostility, and President Eisenhower issued a statement of hope that he would be received courteously. Public officials did not respond, but I felt it was important for him to see at first-hand our industrial capacity and efficiency. He visited the automotive plants and The Detroit Edison Company. While here, he asked me to make a second visit, this time to Armenia and Siberia.

Another team was gathered from a E.E.I. and we made an extensive mission and prepared a second report. This was at the time Vice President Nixon had his much publicized kitchen debate with Premier Kruschev, but we were too busy trying to understand the Soviet energy situation to get involved in politics. We were there to help the Russian people and to learn how to help the American people.

Then in 1961, I returned to Moscow with the U.S. Committee of the World Energy Conference. Mikoyan invited my wife and me to his office and we spent an hour with him. He talked about the economic development in his country, and I told him about the imbalances in the world and within the Soviet Union. I said to him: "Mr. Mikoyan, you have problems similar to other countries. You have two and a half times the land mass of the United States, but half of Siberia is covered with permafrost. In that great area are 80 percent of the natural resources but only 20 percent of the population."

He had reservations about my report on what had been accomplished in the power industry in the United States, but finally he said, "Mr. Cisler if you tell me this is so I will believe it, but if your State Department tells me I will not believe it."

Since I did not want him to learn of it later in the press, I told him I was on my way to India on another energy mission.

"That is good," he said. "That is good. You should help the people there."

Mikoyan was one of the old revolutionaries who survived many changes and in 1964 became president of the U.S.S.R. He was a moderate man who provided a stabilizing influence. His son, Sergo, who became a scholarly economist, visited Detroit. I gave him a short wave radio and tuned it to a Moscow broadcast. "This is your country we are hearing," I said. "The United States does not interfere with your broadcasts to the American people, but when you get home the jamming of our broadcasts will prevent you from listening to us."

Mikoyan's grandson and great granddaughter have subsequently visited me in the United States. I came to know them well, and I feel that this continuity over four generations of intelligent people is a good thing for both our countries and the world.

I visited the Soviet Union several times in the 1960s and 1970s, and their engineers have studied the electric energy systems of the United States. Because of the Cold War, the relationship has not been an easy one, but communications on technological matters have remained open. They are excellent hosts. On a recent trip, I reached Moscow on Thanksgiving Day and had taken with me several mince and pumpkin pies. Pyotr S. Neporozhny, the minister for power and electrification, met me and then arranged a warm and friendly party of his own thanksgiving to celebrate the occasion. I cut the pies into equal pieces so the girls waiting on the table could have their share.

The American power industry has learned much from the U.S.S.R. about planning. The Russians come at this problem the other way around from us, and it is worthwhile to be able to observe the differences. In the United States, we start with the wishes of the individual home dweller for electricity and add them together with the wishes of individual governmental agencies, businesses and industries. This is the demand approach and, because of the abundant resources with which we have been blessed and take for granted, we have been able to build the plant to satisfy the demand. This is also largely true of the whole free market world. In the Soviet Union, the government starts with the total resources that are available and the planners then allocate to the energy field what they believe to be its proper share. This is the supply approach, and it is substantially the same in the whole controlled market world and in those nations developing into industrialism. If the time should come when the United States would not have primary energy adequate to our need then we, too, would have to consider how to allocate scarce resources.

By 1960, the scars of war had been pretty well healed in Europe and Japan, economic stability had been achieved and life there returned to normal. The United States Agency for International Development (USAID) began to concentrate on the developing countries, and my activities with power surveys and other economic studies increased.

The day I left Mikoyan, I flew to Tashkent and then to New Delhi where Tyler Wood was head of the U.S. Economic Mission. This was my first trip to India, although I had worked with Wood in Korea and Japan. After a few days of reviewing the situation, I recommended a power and a total energy survey. The following year I arrived in New Delhi at 5 A.M. Christmas Day and met Prime Minister Jawaharlal Nehru who approved going ahead with the comprehensive study. This was a monumental task requiring two and a half years to complete.

M.R. Sachdev, secretary of the ministry of irrigation and power, chaired the committee, and Professor Austin Robinson and I were co-chairmen. In addition to a number of high ranking members of various departments of the Government of India, Jacques Desrousseaux, director of economic studies of the French Coal Industry, joined the group. He had worked on the Robinson Report for OEEC. And my old friend, Louis de Heem, director of the Center for the Study of Nuclear Energy of Belgium, with whom I had worked in Brussels during the war, and then when I was with the military government in Berlin and still later with the Marshall Plan, also joined us. The working group was directed by Arthur E. Bush and Harry Tauber, both of The Detroit Edison Company and both veterans of a study of Tunisia. Robinson wrote most of the final draft, and the *Report of the Energy Survey of India* was the first outside study accepted as an official document by the Indian government. In January, 1975, Queen Elizabeth knighted him for his work in Europe and India. He was secretary of the Royal Economic Society and joint editor of the *Economic Journal* for twenty-five years after the war, and Sir Austin is still actively pursuing his economic studies. I valued his dry wit and common sense.

Sachdev was the most influential man on the committee. He was a graduate of Robinson's college at Cambridge University and had been trained in the tradition of the Indian Civil Service. After World War II, he had been involved in the partition of India and Pakistan and the return of the refugees. He understood his people and the very difficult conditions under which millions of them lived. The difficulties of my boyhood were nothing compared to those faced in the developing nations. We had anthracite coal. The Indians had cattle dung, gathered carefully at dawn by the women, patted into flat discs and then left to dry in the sun. The forests were rapidly being denuded for essential firewood. What coal existed was of low heating quality and was concentrated in the northeast. The limited supplies of oil were in pockets in the northeast and northwest. The hydraulic potential was distributed a little better although almost half was in the north toward the Himalaya Mountains. The problem of generating enough electricity to serve the agricultural and industrial needs was matched by the difficulties of transmitting it to almost 600,000 towns and villages throughout the subcontinent.

In the midst of our studies, the Indian army occupied Portuguese Goa, and Sachdev was named lieutenant governor of this newly acquired territory. He died there of a heart attack, and the country lost one of its most capable public servants and an extraordinarily fine human being.

At about the same time, Dr. K.L. Rao, whose friendship I

have valued ever since, was appointed minister of irrigation and power, and he continued the study. The conclusions pointed out that India had imbalances that would require years to correct. Animal and vegetable refuse were needed for fertilizer rather than fuel. Beyond what could be supplied by hydro resources and indigenous coal, nuclear energy seemed the only good answer for the generation of electricity to pump water for irrigation as well as for industrial production. The population had increased from 440 to 550 million people in the decade of the 1960s—one million new souls a month. Despite the high yield of miracle seeds and industrial fertilizer which produced a so-called Green Revolution, there was actually less food produced per capita at the end of the decade than at the start. The 1:7 ratio between energy and all other capital investment was as valid in India as in Europe.

Nehru kept close to the project, and at the end of the study, I presented him with a biography of Thomas Alva Edison. "This is an important book," he said. "My grandson must read it."

India brought me face to face with the difficulties of those nations that were emerging from a largely agricultural to a more industrial economy. Considerable time was expended on the political question of whether a third world could emerge between the United States of America with its allies and the Union of Soviet Socialist Republics with its allies. The Indian national atmosphere was dominated by more regional problems than I had ever seen before. And the economic life of the Indian people was dominated by severe imbalances. The country really did need its own Thomas Alva Edison.

Whereas India had a large population and few energy resources, Iran had just the reverse and presented a quite different economic pattern. My first contact with that country was made under the lend-lease agreements to ship equipment and military supplies to the U.S.S.R. by way of the Persian Gulf. Teheran grew very rapidly after the war but there were thirty-three small electric utility companies serving the community. The service was very bad. Mansoor Rouhani was minister of agriculture. He came to visit me in Detroit, and the USAID Mission agreed to support a study. This resulted in the development of a national program by regions. Teheran Electric Company serves the largest region, but there are others in Isfahan, Shiraz and the major cities. In 1967, I first met Shah Mohammed Pahlevi and presented him with a copy of the *Urban Detroit Area Study*.

119

In the middle seventies, I am still consulting with the Iranian government on energy matters. Through Overseas Advisory Associates, Inc., I am working on the development of nuclear generation for the country and on training their engineers and operators. Even though Iran has very large oil reserves, they are a finite amount. The Shah wants to preserve this valuable resource for future generations of his countrymen. With oil's long-term value in the transportation and petrochemical industries, he does not want to burn an unnecessary amount to generate electricity which can be produced by uranium.

One of the most recent energy studies and reconstruction projects has been in Vietnam. Once again we had Detroit Edison engineers there at the request of the United States government and Harry Tauber led the team. Throughout the war the construction of generating plants, transmission and distribution lines went forward steadily. Although American planes bombed power plants around Hanoi, particularly in the last days of the war in December, 1972, the North Vietnamese did not attack installations in the south. Electricity is a scarce and essential commodity in developing nations on which they place great value. It is a public resource, seldom destroyed. Up to this date the *National Power Survey of the Republic of Vietnam* is the only complete study and set of recommendations that have been made for the continuing development of Southeast Asia. The plans that were developed up to the year 1985 will also bear fruit in Cambodia, Thailand, Laos and North Vietnam.

In World War II, I was distressed by how much agony was inflicted on people through death and wounds and through the destruction of their homes and means of gaining a livelihood. Under the Marshall Plan and subsequent efforts, I learned first-hand how laborious is the reconstruction of merely the physical elements. Despite this, I accepted the political judgments that took the United States into the Korean conflict. But I questioned the Vietnam War which not only brought great destruction to that divided country but also seriously divided the American people. The damage to both the spiritual and physical systems have been great, and the reconstruction at home and abroad will be long and difficult.

Technical assistance programs like the ones described here were carried on in many nations as part of that civic responsibility felt by American businessmen and engineers. I

120

myself have been involved in projects on every continent, and Detroit Edison people have participated in studies of more than twenty countries. We have broadened our understanding of both technical matters and people. We have given and we have received knowledge. From time to time, there has been criticism of all this activity. Some have asked why one utility, tied to southeastern Michigan, was involved in so many parts of the world.

I was flying home from Europe in the 1950s when it occurred to me to apply knowledge being gained there, through our total energy studies, to the state of Michigan. As a result we have made two studies here and have strengthened interconnections among our neighboring systems for the economic sharing of power.

I was returning from Pakistan when I first considered that the comprehensive land-use studies being made by Doxiadis in that country could be applied to the service area of The Detroit Edison Company. The result was the *Emergence and Growth of an Urban System* that stretched our minds beyond our earlier ability to conceptualize. It identified the realities with which we live on a daily basis, and I am sure it will be a fundamental planning document for many years.

I was coming home from India when the idea of the Monroe power plant came to me. As I considered the lack of primary fuel in the country and thought about the rising demand for oil in Europe and Japan, I became convinced that Detroit Edison should generate its base load from coal. Within the borders of United States is 40 percent of the proven coal reserves of the world, and much of it is within 500 miles of Detroit. For a few years, it looked as if rapidly rising environmental demands might force a political solution that would counter this economic decision. On the east and west coasts in particular, and in other urbanized areas some utility companies were required to convert old plants and build new ones to use oil. By the fall of 1973, however, the realities of the oil economies overtook the United States as informed people had known it would.

During the thirty years following the end of World War II, some of the largest multinational corporations were developing world headquarters in Detroit and in our service area. I was a member of the board of Burroughs Corporation and chairman of the board of Fruehauf Corporation. Several of the local banks established agencies in other countries. These banks were our customers and the world was their customer. I felt it was quite appropriate as well as sound business for The Detroit Edison Company to have an international perspective.

CHAPTER VII

NATIONAL AND STATE AFFAIRS

Whereas political ideologies tended to set the course of affairs in those international activities with which I dealt, economic realities dominated the domestic scene in the three decades following World War II. There were serious anxieties, of course, about the intentions of the U.S.S.R. and later China, and there was debate about whether socialist doctrine was an internal threat to our political system. Americans were primarily concerned, however, with improving their living standards following the austerity of the war and depression years. The major parts of the American free enterprise system—agriculture, business, industry, labor, professions, and services—were interested in steady economic improvement.

My own political activity was typical of that of many Americans. I have cherished the right to disagree with our presidents on a number of issues, and I have gladly served the country under them. Many people would classify me as a Republican because I represent the management of big business, and I probably tend to think of myself as one. In actual practice, however, I am an independent. In the primary elections I sometimes vote the Democratic ticket to help place the strongest people in contention. I never vote for those I consider the weakest in order to enhance the odds that the Republicans will win. In the regular elections, I am a confirmed ticket-splitter. I seek the most capable individual to support, regardless of party.

My experience around the world has led me to conclude that development in the United States does not fit the ideological patterns of much of Europe and Asia. The role of the courts in our constitutional history, and the relatively open access to legal remedy for grievances have apparently been adequate to our needs for change. By and large, we are disciplined in the law rather than in political parties. There were several occasions in the 1950s when I might have been given a cabinet appointment had I been interested. Instead of spending my time in partisan politics, however, I preferred to try bringing Republicans and Democrats and business and laboring people together in cooperative endeavors.

Before I came to Detroit, the state of Michigan had been strongly Republican and Governor Wilber M. Brucker had guided it early in the depression years. I met Brucker later in Germany during the period of military government, and I worked with him again when he was secretary of the army in the first Eisenhower administration. In 1948, however, the Democratic party came to power under G. Mennen Williams. After returning from the Navy, Governor Williams put together enough support throughout the state to hold his position for twelve years, an unprecedented tenure. I have voted for him and for his opponent. On a number of occasions, he asked my help in matters related to the economy of Michigan and I always gave him my fullest support. It is enough to say that at home as well as abroad I have been much more involved in government than in politics. Government is the business of the political system. It impinges on an increasing number of activities in our lives, both public and private. It is in this area where imbalances cause trouble and where we seek the finest adjustment between individual and private good with social and public good.

When I became president of Detroit Edison in 1951, I found myself participating in a broad range of national and state affairs which I tried to keep related to the primary concerns of the company. In my mind, like the international activities, they were largely in the economic and social channels of organization.

Detroit in the 1950s

When I came to Detroit in 1945, it was to a city that had played such a leading industrial role during the war that President Roosevelt referred to it as the arsenal of democracy. The Korean conflict raised the World War specter of scarcity of materials again, and the Department of Interior established the Defense Electric Power Administration and invited thirty executives of private and public utilities to participate. They were concerned about an adequate supply of electricity as well as generating equipment and material. The semi-annual power surveys of the Edison Electric Institute formed a good starting point for the executives' work, and generating capacity was increased 25 percent in the next three years.

About the same time the Commission on the Defenses of the North American Continent was formed by Robert Lovett, secretary of defense under Truman and later continued by Charles Wilson who left the presidency of General Motors to take over the defense position under Eisenhower. I was able to bring the early warning radar studies into Michigan as part of the DEW line against possible trans-polar air attacks. It was a serious matter and related directly to the siting and size of the company's power plants. For a time, it was considered strategically desirable to limit the plants to 500 mgw in size, but as fears of war receded and the efficiency of larger machines became greater, we chose the economies of scale.

Governor Williams was greatly concerned about the impact of the Korean War on the economy of Michigan. He named a Commission on Productivity which included Douglas Fraser and other labor leaders as well as people from business and government. Korea was taking supplies away from the automobile industry. He asked me to chair this commission and, of course, I agreed to help in any way I could. He was in touch with the people of the state and I respected his ability.

The productivity commission met once a month to analyze and monitor what was happening. I tried to put information together into meaningful systems in order to meet actual situations as they arose. Governor Williams would sometimes phone me at 11 P.M. to talk about his concerns. I had a high opinion of him, and I did have good communication with Washington bureaus and their allocations people. Fortunately, the conflict did not expand into full scale war with China and conditions began to ease within a year. Industrial production had grown to such remarkable capacity that the United States could absorb all the dislocations of this rather major event, although the accompanying inflation dampened the real improvement going on in people's lives. Gradually Detroit reverted to its pre-war image and was again known around the world as a center for transportation and economic opportunity. People who were frequently unfamiliar with the name of Michigan and occasionally with the city of Washington responded to the mention of Detroit.

Although The Detroit Edison Company serves customers in all or part of thirteen counties in southeastern Michigan, it began in the city of Detroit and still has its largest load in Wayne county. Until quite recently, it had its largest investment in power plants and other facilities in the city. In some years it has been the largest taxpayer and has long been among the three largest. As a result there has been a mutuality of interest between the company and the city that is clearly important to both parties.

As soon as I became settled, I got acquainted with the community through Prentiss Brown and his broad involvement in public affairs. There was a small, informal group of religious, labor and business leaders who used to meet periodically to talk about the city. They invited me to come in, and I thought it was a wonderful opportunity. I came to know about the inner aspects of unions, blacks, middle-class whites and professional people, the difficulties they faced, and their aspirations for a better order of things. Many of them shared common experiences during the war which shaped their hopes for a more rewarding way of life. Stanley Kresge, who was a dedicated Methodist layman and who was practicing one of the great merchandising concepts of low cost distribution of goods, was in the group. There were such men as the Reverend A.A. Banks, who was pastor of one of the large black churches; John Dancy, president of the Urban League; Allen B. Crow, president of the Economic Club of Detroit; John Danhof, president of the Detroit College of law, and John Coleman, chairman of the Burroughs Corporation.

In 1951, the city celebrated the 250th year since Cadillac put his canoes ashore and founded the first of a continous line of settlements on the Detroit River. The celebration generated considerable civic spirit and a number of committees were organized to support events associated with the anniversary. A Capital Gifts Committee headed by Frederick C. Matthaei raised $9 million. One million dollars was given to Wayne University to build a Community Arts Auditorium and so expand the cultural center of the city. The remainder went to build a convention and sports center named after Mayor Albert Cobo; it strengthened facilities in the central business district.

Cobo encouraged business leaders to continue the momentum provided by the birthday. He was quiet but effective. He had a vision of the future Detroit and he was a good administrator. The Detroit Edison Company participated and, in addition, I was chairman of the International Visitors Committee set up to welcome participants in the celebration. Following the birthday year, this committee remained active as the Office of International Relations, and in the next decade thirty-eight heads of state plus other political leaders from twenty-five countries visited on official business as guests of the city. Queen Juliana of the Netherlands, who spent two days here, was among the most socially prominent guests, and Detroit gave her a warm welcome. By 1970, the functions had been picked up by other organizations and with the retirement of Mrs. Ruth E. Stevens, the original vice-president, the committee was officially discontinued.

After the Capital Gifts Committee completed its fund-raising goal, the Detroit Conference on Civic Development was organized in 1954, and I became one of the co-chairmen. Out of this group a sub-unit was formed called Detroit Tomorrow, which was asked by Mayor Cobo to work with various departments of city government. Once again I found myself deeply involved in civic activity, a part of my long and continuing interest in working for a more complete understanding between the public and private sectors in the country as a whole and around the world. This group ran out of forward direction and in 1958, when Louis Miriani became mayor, it was turned into the Committee on Economic Growth because the business downturn in 1957-58 affected Michigan more than most other states and Detroit more than most other cities.

I worked hard during the 1950s to build up momentum for improvement. There were scattered successes that were encouraging, but basic elements were missing. We did not have a good plan for the total investment that would be necessary as we looked at the anticipated development. What were the total capital requirements of the state? We were a great productive force and we needed capital. As my own awareness of worldwide economic forces increased, I knew that the state, just as developing nations, needed one-eighth for energy development for every seven-eighths going into all other capital requirements.

I remember one evening in the early 1950s when I laid out the substance of this thesis to a small group. Cobo was present along with Joseph Dodge, just back from advising General MacArthur in Japan, and Charles T. Fisher, Jr., president of the National Bank of Detroit. I called on my experience as a planning and construction engineer in New Jersey as well as on years with large governmental systems during the war. The strength of systems was that they assembled resources and applied them toward a stated purpose. The problem was that, like chains, they were no stronger than the weakest component. Each of the other men contributed to the discussion out of his special knowledge. We agreed that we had to mobilize the capital resources and get some momentum started.

I was very busy and for a period of three years I never went out of the country. But I had people in many parts of the world and I had an office in Washington. Much of the planning and many of the meetings took place on weekends, and this regimen became my substitute for recreation.

Despite the stop-go economic conditions of the decade, the momentum did build and the various committees did develop a sense of citizenship. People were better fed and better housed than ever before. There were more transportation alternatives, more appliances, more leisure time and more discretionary income. All of this was as good for The Detroit Edison Company as it was for the city and the whole service area.

Between 1951 and 1960, the number of customers of all classes increased by 27 percent and the number of kilowatt hours sold by 75 percent. As a measure of increasing affluence, the average residential customer used 160 percent more electricity. The company's construction program gave it the capability of producing almost twice the number of kilowatt hours. We had achieved a 20 percent increase in the efficiency of the burning of coal, employed 16 percent fewer people and paid an average wage 60 percent higher than in 1951. These improvements, which resulted from the increased size and improved engineering design of plants, were passed along to consumers in the form of rate reductions.

In addition to my involvement with the economic life of the city, I also participated in the developing cultural life. For many years, the Ford Motor Company had supported the Detroit Symphony Orchestra. Then Ford built a new auditorium on the riverfront, and the responsibility for operating costs passed to a broad group of citizens under the dedicated chairmanship of Robert B. Semple, president of the Wyandotte Chemical Corporation. I served as a member of the executive committee.

Although the Detroit Institute of Arts was a city institution, it had been supported since the 1920s by wealthy families. To draw a line separating public from private support, a special corporation was set up called the Founders Society. Money raised by this group was expended principally to purchase works of art, and it was chaired, usually, by a senior officer of a major corporation. For many years William M. Day, president of the Michigan Bell Telephone Company, filled the role. Succeeding mayors also sought corporate leaders to be city commissioners. Lee Hills, president and publisher of the *Detroit Free Press*, chaired the arts commission for many years.

The Detroit Public Library also developed a similar kind of public-private support as did the Detroit Historical Commission. Participation in community affairs, beyond the economic benefits contributed by corporations, represented a cautious change of attitude on the part of businessmen. But it was one I approved of and tried to encourage. My whole experience had demonstrated that people were best served when government and private interests worked together.

Detroit had been a pioneer in the united way of raising private contributions for social services. Businessmen gradually learned that they could help prevent duplication of agencies, reduce administrative costs and deliver more help per dollar if they went at the problem cooperatively. They also found they gained a better understanding of the needs of people for social services and so they raised more money. The United Foundation drive became the largest in the country and collected about $35 million a year. Once every five years, there was an additional capital drive which went largely to build private medical facilities.

Although I was an active member of the Engineering Society of Detroit and served a term as president, the private educational venture to which I gave most personal attention was the Detroit Economic Club. Prentiss Brown had been elected chairman in 1952 and, when he retired in 1954, I succeeded him for twenty years. The club had been started in the depths of the Depression when Allen Crow brought a group of people together at lunch time to talk with one another and to hear experienced people discuss serious issues. The objectives were very congenial to my own sense that business leaders required a wide mix of information and points of view to function successfully in growing systems.

The club's original format has continued to the present day. Anywhere from 500-1,000 people gather every Monday in the central business district during what is, roughly, the academic year, and usually some member will host a group of students and teachers. The guests may come from high school or from private or public colleges, and it gives them an opportunity to join with the business community in the important matter of exchanging ideas. In 1973, women were admitted to membership in recognition of their increasing leadership roles in our society.

Engineering and Education

I carried on my interests in engineering and engineering education in a number of ways throughout the 1950s and 1960s. Because of my early contacts at Cornell, I remained active in the American Society of Mechanical Engineers and served on a number of committees before becoming president in 1960-61. In 1962, I was greatly honored by being selected to receive the twenty-third award of the Hoover Medal which commemorates the civic and humanitarian achievements of President Hoover. Receipt of this award linked my name to those of Vannevar Bush, Karl T. Compton, Alfred P. Sloan, Jr., Charles F. Kettering and Dwight D. Eisenhower. The Engineers Joint Council, which coordinated the common interests of all professional engineering societies, chose me as president in 1964-65. There was a good deal of discussion within the council, at this time, about the status of the profession. The National Academy of Science was an old and distinguished organization that had been recognized by President Lincoln during the Civil War. The National Science Foundation was a recent group which was supported by an annual appropriation of over $300 million from Congress. The space age was five years old, and the race to beat the Russians to the moon had been publically announced. Physics was once again the queen of the sciences, ably supported by mathematics. Congress voted billions of dollars to support the National Aeronautic and Space Administration (NASA). Engineering was thought to be less important as was revealed by smaller college enrollments. So in 1964, the council founded the National Academy of Engineering as a parallel organization to the National Academy of Science, and I was one of the twenty-five founding members. Its chief goals were to make available to Congress and the government the profession's talents, and to give engineers national and international recognition.

One of my most interesting activities occurred in 1959 and 1960 when I was called upon by Secretary of Defense Neil

McElroy and by the administrator of the National Aeronautics and Space Administration, Dr. Keith Glennan, to advise them on the matter of space travel. The Soviet Union had launched Sputnik I on October 4, 1957, and the beep it emitted as it circled the earth, the first man-made instrument to enter outer space, was heard as a challenge by almost every American scientist and engineer. President Kennedy was not yet in office and had not yet blown the whistle to start the race to the moon, but the contestants were warming up. Dr. Glennan had been a member of the initial Atomic Energy Commission which had called upon me in 1946. Now he was with NASA. These two men knew that the United States had the capability of putting an object into space. What we did not have was the capacity to put a man into space and then bring him safely back to earth. It was in the study of this problem that I was asked to participate and to recommend a managerial structure that would ensure clear-cut control over ground facilities and space flight, efficient use of national resources and effective action to fulfill objectives of national policy.

For the first six months I visited areas in the Atlantic, the Pacific and the Arctic oceans to look at possible launching and return sites. I also began to consider the complex communications systems that would be needed to control the missile systems and monitor the astronauts. It was another exciting frontier period like the development of nuclear technology. The concept of the planet earth as a spaceship developed in people's minds, and it produced a spirit of adventure very much like the one in the early Renaissance when the travels of Columbus and the first transatlantic navigators established the mental image of the earth as a ball. When men were released from the dread of sailing off the edge of a flat saucer, commerce followed the new knowledge as voyage followed voyage and Europeans turned outward from their small kingdoms. In a few years, the millennium we call the Dark and Middle Ages ended, and modern history began.

My recommendation was that NASA centralize all communications in a single center, and once the orbits of the vessels were charted, Houston became a logical site. It was from this spot that we chose to monitor the earth rotating on its axis every twenty-four hours as it travels at a speed of some 67,000 miles an hour in a circumferential passage around the sun of some 584 million miles during the course of a year. I am glad to be alive in a day when we can enjoy the sight of the earth in its ordered passage through space.

One of the areas of social responsibility in which I have had the greatest interest has been education. I am very grateful to the teachers who did so much to help me form my own life. Beyond that, they taught me in such a way that I came to enjoy intellectual discipline and the process of learning. But most of all, I became convinced by my experiences from boyhood on that the application of the mind is the only way to solve the problem of work. It is the only solution to productivity. It is essential that an urban, industrial and democratic society have a highly educated citizenry able to understand, make reasoned judgments and provide creative ways to improve the quality of life. To celebrate its fiftieth birthday in 1953, The Detroit Edison Company introduced a twenty-year program of state-wide college scholarships. Although most of my available time was spent on higher education, I tried to advance the teaching of science in high schools in a number of ways, especially through the Thomas Alva Edison Foundation. I succeeded Charles F. (Boss) Kettering as president and each year the Foundation put on a number of regional conferences for high school teachers. Once a year the Foundation celebrated Edison's birthday both in the United States and in one foreign country, and several students and teachers were given the experience of joining with their counterparts abroad.

After graduating from Cornell and throughout the 1930s and 1940s, I remained active with the engineering alumni association and kept my contacts with the faculty. I was honored to be named a trustee of the university and served in this position for thirteen years, becoming chairman of the Executive Committee.

Inasmuch as The University of Michigan is one of the leading institutions of higher education in the United States, I turned to it for help in many ways and also attempted to support its programs. Dean E. Blythe Stason did pioneer work on the laws dealing with atomic energy. In order to strengthen the university's capability in nuclear engineering, Detroit Edison joined a number of businesses to help fund the Phoenix Project which brought the first teaching reactor to the campus. Harvey Wagner, who was executive vice president of Detroit Edison and one of the most experienced men in nuclear engineering, was a graduate of the university and the company supported the Harvey Wagner Professorship of Electrical Engineering.

The University of Michigan's Graduate School of Business Administration organized the first energy survey of a single state. *Energy and the Michigan Economy* was published in 1967 and revealed that the state imported 95 percent of the total energy it used. This amounted to about 10 percent of the Michigan gross product or, roughly, two billion dollars a year. The book was the combined effort of a number of industrial experts and members of the business school faculty including Paul W. McCracken who later became chairman of President Nixon's Council of Economic Advisers and still later a member of The Detroit Edison board of directors and an adviser to President Ford. The idea for the study came in 1962 when the United States government was applying the results of European reconstruction to other countries around the world. Since Michigan was an advanced industrial state with an economic base many times larger than many of the countries already surveyed, it seemed only prudent to apply some of what we had learned about planning principles.

Economics and the Environment

In an age that has produced specialization and esteems professionalism highly, it is easy to consider economic and commercial matters as being valuable for their own sakes, although I have always emphasized people. Like baseball statistics, the Gross National Product (GNP) is a very useful yardstick for measuring performance, but statistics are not the same as the game itself. The purpose of our business enterprise is to make the life of each individual a little richer. At present, there are almost four billion people on the spaceship earth, and within the last quarter of this century the number could grow to six or seven billion. It is sobering to contemplate how this situation is to be met at all, let alone how the lot of each individual will be improved. Along with most people of my generation, I have accepted growth as obviously necessary and certainly desirable. The lives of people have improved immeasurably. The only questions that require answers are how much growth and how fast?

As early as 1948, I was asked to talk about the problems of long-range growth, the need for energy as a key part of this growth, and the economics of fossil fuels and especially of nuclear energy. The United States had 6 percent of the population of the world and used 35 percent of the energy in order to enjoy its urban industrial life. By the end of the 1950s, it was clear to many of us that the demand on fossil fuels was becoming too great. With good management of the resources of the earth, there were enough for 500 to 1,000 years. But the strain of extracting and transporting them was already being felt in many countries. The coal miners of Europe could no longer compete economically with oil. In the United States, I met John L. Lewis of the United Mine Workers to explain how atomic energy would only supplement coal and would not supplant it in the foreseeable future. The Detroit Edison Company also published a booklet on coal with the endorsement of the United Mine Workers.

In one form or another, then, the economics of energy occupied much of my attention at home as well as abroad. The anticipated crisis could be averted if proper decisions were made far enough in advance. There was no shortage of energy. There were shortages of information in some fields and of good management in others. Governmental agencies competed with one another but often worked at cross purposes. It rapidly became important to try to understand the nature of these conflicts and to help resolve them.

In order to reach a better understanding of the complexities and to share with others in the formulation of solutions, I began to participate in a number of national activities. I became a director of the American Management Association and gave several talks on the management of the power industry to engineering societies and universities. In the middle fifties, President Eisenhower sought a group which could advise him periodically about the condition of the nation's business activity and he turned to the Business Advisory Council in the Department of Commerce which later became the Business Council. I have been a member since 1954.

A typical meeting would be held at Hot Springs in Virginia and political and governmental leaders would attend. We talked in general about the economic health of the country. Since the business downturn in 1957-59 hit the state of Michigan with particular force, I was most concerned with that activity.

President Johnson formed a Task Force for the Office of Economic Opportunity in the war against poverty. This was the kind of war of which I approved, and I served as Chairman of the Business Leadership Advisory Council from 1965-69. These contacts with Washington were especially valuable to

Mayor Jerome P. Cavanagh as the city of Detroit tried to cope with its growing urban problems.

During the twenty years from 1946 to 1966, the GNP grew at a rather steady rate of three percent and all forms of energy kept pace. This was healthy growth which provided a doubling of the economic standard of living in twenty-five years. Electric power grew at the same time at a rate of seven percent, doubling every nine years, and this also was manageable. It reflected a larger penetration of the energy market brought on by increased use of sophisticated machines. In the home, the electric fan was replaced by air conditioning. In industry, coking ovens were replaced by electric furnaces in the making of high grade steel. Individual machines grew more productive and were put into complex systems that demanded an upgraded form of energy like electricity. The human body has only limited capacity to work, but machines theoretically have unlimited productivity.

The use of electricity was also increased by environmental forces. The Clean Air Act of 1964 was amended in 1967 and then made seriously restrictive in 1970. It began to put a strain on electric utilities to provide the clean energy necessary to meet the law. In addition to the task of enhancing the environment within homes and places of work, the industry now was called upon to improve the quality of the surrounding air and water. It seemed inevitable that something close to a seven percent growth would have to continue if the industry was to improve both economic welfare and the physical environment. There were trade-offs in all this. The amenities of life might be improved at considerable cost to the necessities. Once again, it was a matter of balance. Despite the popular protest against technology, the significant question was the cost of approaching zero pollution, and the problem was not seriously addressed until inflation began to undermine our standard of living.

The Detroit Edison Company has always been a leader in environmental matters. As early as 1924, it installed the first mechanical precipitators to take some of the particulate out of the smoke. Over the years, improved and electrostatic precipitators were added until they were removing from 98 to 99.6 percent of the solid matter. After 1951, when I became president, every new power plant was designed to meet company standards that were superior to those of the state or federal government until 1967. We also built higher stacks as the plants grew larger in order to dissipate the sulfur dioxide and other gases found in coal. At the latest Monroe plant the stacks are 800 feet tall, higher than any building in Michigan. By the best figures obtainable from government sources, the air now is cleaner than it was in 1948 when most Detroit homes were heated by hand-fired coal furnaces. Even so, the growth of technology had a severe impact on the environment. There were more automobiles and more industry each year.

Within the company, we appointed an environmental officer and in 1969 created a new division in the Engineering Research Department. Over the next several years, we spent hundreds of millions of dollars for environmental equipment. The cost of the additional energy required to run the equipment soon exceeded in each year the total investment of The Detroit Edison Company in the Fermi demonstration project. As a private entrepreneur, I considered the R&D costs for the Fermi technology great enough to require a consortium of companies to finance it, and the expenditures became a matter of high public concern. The much larger and continuing environmental costs were required by government without any real public awareness. The hurry-up legislation served an immediate political purpose but at a severe cost to rate-payers. It may or may not be fair to say that the critical energy situation in the United States was legislated into existence, but it can safely be said that the wide variety of conflicting laws and bureaucratic actions have certainly made a significant contribution to it.

Detroit in the 1960s

Time is a very important factor on either a rising or a falling curve. Unless there is sufficient momentum an upswing will stop moving forward, slide over the peak and start down. In many ways the United States, Michigan and Detroit were improving in the twenty years following the end of World War II, but in some ways imbalances were beginning to appear.

I have always been mindful of the social, the human, side of everything I was engaged in. The sense of these values was engendered early by my mother, and I carried them through all subsequent activities. As I look back over my life, I realize that few things have moved me more forcefully than to see someone in trouble. I react immediately to individual, corporate or national distress. Among my vivid memories is a little girl in Palermo crying her heart out because her parents had been killed. My recollections of Berlin are troubled by the arms and legs that could be seen sticking out of the debris of war. I still recall a German farmer during the first days of the military government. His house and all his possessions had been destroyed, and he was driving a horse down the road with his children in the wagon and a cow tied behind.

Corporations also need a sense of humanity. In our free enterprise system, profits are necessary to give the shareholders a just return for providing the capital and undertaking risk. Only profitable firms can continue to provide good service to customers and to pay employees a respectable wage. But in my judgment, corporations do not exist only to make money; they exist to serve people.

It is necessary to have this sense of social values inside the corporation before it can manifest itself outside. This is the matrix for employee relations. Good morale requires more than just meeting legal and financial obligations, important as it is to have these spelled out in labor contracts and employment agreements. Management cannot decree character in an organization any more than politicians can legislate it. When social awareness exists inside the organization, however, it then spills over into a total sense of corporate responsibility to society.

It was in 1961, when I stopped over in West Pakistan, that I saw Constantinos Doxiadis at work again. He had returned to the private practice of architecture in Athens following his work with the Greek government and the Marshall Plan, and one of his projects was to build a new city of Karanga just outside Karachi. He had received a grant from the Ford Foundation and was building homes again and schools for people who were living in abject poverty. Flying back over the Atlantic I thought if we could accomplish this much good in Pakistan with some intelligent planning, why not try it in Detroit.

The Federal Highway Act, passed during the Eisenhower administration, had done a great deal for intercontinental roads and made possible the construction of expressways in the cities. These roads had already been started in Detroit to prevent the downtown area from being completely choked with traffic. With more federal funds, the process was speeded and soon Detroit had thousands of persons displaced by road construction. The expressways cut through natural communities that had been organized around schools and stores and disrupted human lives socially and economically. As I drove to work each day, I passed through the growing blight that spread out ahead of the expressways. Mayor Cobo had spoken many times of his pride in Detroit as a city of homes. There were few apartment buildings and an unusually large percentage of home ownership. Now people were forced out, usually at a financial loss, and we had created an inner city of displaced persons.

We had begun to partition the city at a time when migration from southern farms, that had really started during the war years, had reached a significant level. These people were both black and white and they lived in poverty, poor health and poor housing. They were ill-equipped for urban living. They brought with them an education which was inadequate for an industrial city. In a very few years the Detroit public school system, once one of the strongest in the country, found itself in deep trouble, trying to provide adequate educational opportunities. So even the future looked bleak for these new residents. I watched the bulldozers tearing down homes and stores day after day and I said to myself, My God, where is this going to end?

All of this set me to thinking hard about solutions. I was aware that the history of Detroit, since the founding of the company in 1903, had been one in which the more affluent and better educated classes moved from the central city into a series of suburban arcs. Successive waves of immigrants occupied the homes they left. Germans and Italians were followed by Poles, Russians and Jews, and later by Arabs and other Middle Eastern groups. As each ethnic community was established, its members soon began to take part in the exodus, dispersing first throughout the city and then the metropolitan area. In the 1950s and 1960s, poor blacks and whites migrated to replace them. Regional growth was occurring, and I began to look at the need to strengthen the transmission system by building additional generating equipment at St. Clair and an entirely new plant at Monroe. Throughout the 1950s, however, people thought of Detroit as the area within the corporate limits of the city.

I became a member of the Metropolitan Fund that had received support from the Ford Foundation to study a larger geographical area. Most of the major corporations had representatives on the board of directors, and the concept of a tri-county metropolis of Wayne, Oakland and Macomb counties began to take shape. In terms of the experience of The Detroit Edison Company this was geographically inadequate, but in the early 1960s it represented advanced thinking. Since I could see no comprehensive land-use plan for the area, and since it was becoming necessary for the company to look at least ten, preferably twenty, years ahead to acquire property, I asked Doxiadis to make a study.

One Sunday in my office we agreed that the only way to approach such problems was with a total concept. I related the ideas I had developed out of the Marshall Plan and Japan and the developing nations on the relation of people to natural resources, to energy and productivity, and finally to planning and management. General Marshall had put the question: How do you help the Europeans to help themselves? Doxiadis had helped develop human settlements in Greece, in eastern Australia and in Pakistan. We discussed the total urban complex and he worked out the methodology of the study. What I wanted was a comprehensive plan for the Detroit area that was like the rebuilding of Europe or Asian countries, that would help Michigan help itself.

Then I started to ask if people would accept such a study. I discussed it with officials and businessmen of Detroit and surrounding suburbs. I worked on this aspect for two years in some hope that others might join as they had in the development of Fermi. In the end, however, I learned that the study would not be resented if Detroit Edison proceeded alone. As a utility, the company had a good public position and people expected us to be advanced in our forward planning. It was recognized that this was more important in a capital intensive industry which, because of the huge construction demands, could not respond quickly to changes in the economic climate. President Clarence B. Hilberry of Wayne State University joined our executive group because the faculty was already knowledgeable about the area and students could receive additional education in regional planning. We anticipated there would be need for many more planners in the future. And Wayne, as a public university, could also help insure political neutrality.

The study got underway formally in 1965, and I will come back to it later in this chapter.

Meanwhile, in 1964, Jerome P. Cavanagh, a political unknown, defeated the incumbent mayor, and even before he took office he talked with me about the great spirit of cooperation created by the 250th birthday celebration under Mayor Cobo. Jerry Cavanagh was young, and with great vigor and charm he began to call the local leadership together again.

Racial uneasiness was growing because of the disruption and slowness of the urban renewal program, and he had inherited a serious financial deficit. He persuaded George Edwards to give up his position on the Michigan Supreme Court to become his police commissioner. Edwards had a long history with the UAW and Detroit Common Council and had an excellent record for justice and concern for people. Cavanagh also persuaded Alfred M. Pelham to leave his position as controller of Wayne State University to undertake similar duties for the city. Pelham was a member of a black family that had lived in Detroit for several generations and had an outstanding record of public service. His father had been controller of Wayne County, and Al Pelham succeeded him and held the position for thirty years. In a matter of a few months, he persuaded the business community to support a city income tax as the most equitable way of removing the debt and insuring financial stability. Then he persuaded the Michigan legislature to approve the first income tax at the city level in the state.

Mayor Cavanagh began to look toward Washington for additional funds from the Kennedy administration. The labor unions, and especially the UAW, were at the height of their influence in the Democratic party. John Kennedy had launched his campaign for presidency by marching down Woodward Avenue on Labor Day, arm in arm with Walter Reuther. Vice President Lyndon Johnson had come to Detroit in 1963 to deliver a key talk on racial equality in celebration of the hundredth anniversary of President Lincoln's Emancipation Proclamation. He also began his 1964 campaign for the presidency by marching down Woodward Avenue on Labor Day.

In 1966, the Great Society programs were being set up in Washington and Jerry Cavanagh was aware of the current political activity leading to the Model Cities Act. He asked Walter Reuther and me to help him. Reuther's social instincts were keen and we said to ourselves: We must be prepared so that when the bill is passed the city can be the first one into Washington with programs that can be funded. Reuther and I had a common meeting ground in international affairs. We had worked together to support the United Nations. He was interested in the labor movement in Europe and Japan. I had come to know his brother, Victor Reuther, under the Marshall Plan because Victor had directed the international relations of the UAW. Now we decided to try to do something with all the urban renewal land that had been bulldozed for the expressways.

We incorporated Metropolitan Detroit Citizens Development Authority (MDCDA) in the fall of 1966 and by the following June we were trying to raise operating funds. Reuther presided over a meeting of about forty people—a number of them from the former Detroit Tomorrow committee—in the board room of the Detroit Bank and Trust Company. As a result, The Detroit Edison Company put up $50,000 and the UAW $100,000, and then it was agreed to bring the organization under the money-raising machinery of the Metropolitan Building Fund. This slowed things down so the UAW made a one million dollar interest free loan and we were finally ready to proceed.

Then in July, 1967, rioting broke out in the city. Because racial conflict had flared in a few other cities it was popularly assumed that this was also the situation in Detroit. I personally believe that this view is not entirely correct. There were certainly tensions between some representatives of the two groups, but much of the disturbance in the first two days took the form of looting stores by both whites and blacks. Then militants with their popular slogan "Burn baby, burn" moved in and soon fires were burning in substantial middle-class neighborhoods that were still integrated, although becoming increasingly occupied by blacks. After five days of civil disorder during which black religious and political leaders were as helpless as white leaders, the National Guard was called in to patrol the city. Detroit settled down to assess the damage, both physical and psychological. The former was considerable. In several largely black neighborhoods, a number of houses were destroyed or damaged and the equity of their owners wiped out. Psychologically, conflict between civil order and disorder took on increasing racial overtones.

The racial antagonisms demanded the first response. Under the direction of Mayor Cavanagh, the New Detroit Committee was organized. Chaired by Joseph Hudson, Jr., president of Detroit's major department store, it consisted of the top public and private leadership of the community. Henry Ford II, James M. Roche of General Motors, and Lynn A. Townsend of Chrysler took active roles. About ten million dollars were quickly raised and emergency remedial action was begun. This was concentrated in three areas: education, jobs and housing.

Over the years, New Detroit took a number of actions to bring about improvement in the public schools. It supported the formation of Wayne County Community College and called on Alfred Pelham to use his great influence with the legislature to get it funded. It supported the decentralization of the system into eight regional districts to bring the resulting school boards closer to the people. State Senator Coleman Young, who later became mayor of the city, took the leadership in this action. New Detroit also helped organize support to increase school taxes. It is still studying ways of upgrading the quality of the education that is being delivered.

Henry Ford II organized and chaired the National Alliance of Businessmen to provide jobs and training in private business. Thousands of people in the Detroit area found employment as companies re-examined job qualifications to see if they were still valid and relevant to the population of job seekers. Ed George, president of Detroit Edison, directed the Detroit effort and organized the employers into a coordinated effort. In a number of instances federal funds were used to underwrite the additional costs of training of the so-called hard-core unemployed.

Since MDCDA was already in existence, New Detroit turned to Walter Reuther and me as co-chairmen of that organization to address the housing question. Although four million dollars was invested in the effort, the problems were vastly greater than the organization could handle in a short period of time. And people expected quick results. We purchased land for building sites and then got caught in the Housing and Urban Development (HUD) syndrome and acquired more than could be immediately used. We studied lower production costs in the housing industry and got caught in the popular appeal of unit construction. But even the most experienced manufacturing companies in the country like Fruehauf could not manage this kind of unit housing on a mass basis. Housing costs, old zoning regulations and labor practices were deeply entrenched and yielded very slowly to change. On top of it all, we hoped to bring new minority contractors into the business.

In all three areas of education, jobs and housing, public expectation ran ahead of reality. When quick solutions to these problems were not found, people again began to use "eggheads" and "do-gooders" as terms of contempt to refer to men's intellect and spirit of compassion.

Even with all these difficulties, I think MDCDA succeeded in opening men's minds to the nature of the problems and to thinking about new approaches. Had Walter Reuther lived, I am sure the organization would still be functioning. He was the key man, and he did not hesitate to speak out against people who were timid or who were in opposition. I represented the more analytical and business-oriented point of view, but we were completely together on goals. We both saw the overwhelming need to rehabilitate the inner city, and I supported him totally.

I recall one occasion when I was in Greece on a mission and learned that Reuther was giving an address in the Hebrew University in Jerusalem. Modern transportation being what it is, I was able to fly there and listen to him and return the same day. Reuther was in good form, he had something important to say, and his audience was responsive. His death in the crash of a small airplane on May 9, 1970, was a personal loss for me as well as for the union people he led for so many years. And it was a great loss for the city of Detroit.

In 1967, then, the two decades of economic growth since the end of World War II began to falter. There were many political causes. The Vietnam conflict had turned sour in people's minds and produced a decline in faith in our political leaders. The income tax machinery had concentrated so much wealth in Washington that people turned there more and more for the help they would have sought from their own resources at another time. Welfare, which was designed to meet the very real needs of the poor, tempted public agencies and private recipients alike to look for a handout to solve problems instead of hard work and even harder thinking. All this economic power in Washington was politically irresistible and people became less mindful of the value of money. There was a real change in the temper of the times and in the attitudes of people. Our normal frailties were extended, and the idea that some inflation was manageable, so a little more would not hurt, was widespread.

Despite all our efforts since the end of the war, very little new venture capital had been attracted to Detroit itself. Until 1964, when the Detroit Bank and Trust Company built a new headquarters, not a single commercial building of any size had been constructed in the central business district since 1929. Following the disturbances of 1967, the business community became more aware of this and basic economic decisions were made to change the pattern of three and a half decades. The difficulties had not developed in three and a half years as many people assumed, and they had not been experienced just by, nor directed solely at, the new migrants from the South.

The Doxiadis project was designed to study a comprehensive solution to the imbalances of population and resources, and it confirmed the long lead time that is necessary to change complex systems. The city had begun to show certain signs of decay as early as 1901 when its business leaders started moving to the suburbs in search of living space. It had happened to Philadelphia in my youth when businessmen bought farms in Gradyville and commuted. Alex Dow had bought a house in Ann Arbor and a farm nearby, and he commuted daily because interurban public transportation was better in 1910 than it was in 1970. For decades the movement away from the city was hidden by the general population expansion into it.

Wherever I have travelled around the world I have observed that people love the land and they want to possess it. They want to own an acre or more and they want to build for themselves and their family their own equivalent of a castle. The city was the place to acquire wealth, and as the middle class expanded decade after decade, the more affluent gradually moved away from crowded conditions for a wide variety of reasons. It is popular to say this represents the rich trying to get away from the poor and this may indeed be part of the explanation. But in terms of actual land development, it is too facile an explanation to help very much.

In Detroit, people moved from lots with a thirty-foot frontage in the old city to ones with fifty feet a little farther out. Some localities toward the boundaries had one hundred-foot lots. Some in the first arc of suburbs had half-acre sites; some in the second, restricted building to one acre of land, and parcels in some of the exurban areas were five to twenty acres. People were willing to expend large sums of money and even greater amounts of time simply to put more space between themselves and their neighbors no matter how wealthy and congenial. As they moved out, the less affluent moved in behind them. By and large, the less affluent, again with many exceptions, were less able to cope with the urbanized industrial culture so the cities lost the time and the commitment of the top leadership except as a place to do business.

It seemed to me that a public utility with an obligation to provide the people of its area with reliable and efficient electrical service at as cheap a rate as possible needed to understand what, indeed, was happening. As the system grew exponentially, the lead time for planning also had to follow a similar curve. A two-year forecast was once adequate. Then we tried to look five years ahead. By 1965, it was clear that we needed a ten-year forecast to make sound business decisions and we did not have one that was good enough. With the help of Doxiadis we set about to do for the urban Detroit area what had been done by the total energy studies in Europe and Asia. In the greater Detroit area, however, a large number of elements needed to be analyzed, the required methodology was more sophisticated and a thirty-year projection was desirable.

The first volume of the study was entitled *Analysis* and it was published in 1966. Some one hundred elements affecting human activity were identified and their interrelationships examined. They included population changes since the 1900 census; the steady encroachment of built-up areas on farmlands; the development of transportation, of industry and other forms of employment; the changes in recreation, air and water quality, and the growth of education and economic well-being. The conclusion reached was that the daily urban experience had already grown far beyond the 137 square miles of the city of Detroit and by the year 2000 would actually encompass 23,000 square miles. Within this area, which included twenty-five counties in Michigan, nine in Ohio and three in Ontario, the daily life of the people would go on. Since the concept of metropolitan Detroit consisting of three counties had just begun to be accepted, this volume pushed men's minds far beyond anything they had yet considered.

The second volume entitled *Future Alternatives,* published the following year, presented a study of alternative solutions to the problems analyzed in Volume I. From a number of models that held promise, one was selected as solving more of the problems of decay and providing more opportunity for healthy growth. This was the development of a twin city in the Port Huron area. Just as the first volume had presented an unfamiliar scale to the developing urban Detroit area, so this optimum solution struck many people as strange and others as unacceptable. It did not seem possible that the metropolis was sprawling out as rapidly as the projections indicated or that all the necessary political and economic decisions could be taken in time.

In the third volume, *A Concept for Future Development,* published in 1970, Doxiadis looked at the most recent evidence and added a detailed examination of Detroit itself. This showed how the unchecked forces of decay had been increased by the expressway program and how the city needed to reestablish a number of sub-communities organized around schools, shopping districts, jobs, recreation and all the things required by people in their daily lives. The data showed there could be as many as twenty-eight of these neighborhoods. The updated evidence also confirmed the rate of urban sprawl. The center of population was moving out from the central business district at a rate of two yards a day, seven days a week. Confusion and decay had already swept over a number of the nearer suburbs and were threatening the next arc. Some of the cities were able to organize sufficiently to maintain their own strength only to find that unplanned development had bypassed and isolated them.

The study revealed several important things that merited more clear thinking than had been applied up to the present. The developing area was already in being, it was much larger than anyone had considered, and it encompassed one international and two state boundaries. The time factor was also much different from what was usually assumed. In the older cities of the world, a hundred years is not too long a time for a cycle of growth and decay to be completed. In 1900, Detroit had a population of about 286,000 people and by 1950 it had grown almost to 1,850,000. After 1953 it began to drop until, by 1960, it was no more than 1,670,000. By 1970, the census figures showed that the population had further declined to 1,511,000 although the greater Detroit area continued to grow.

Cities are large and complex systems, and change in attitude comes slowly. Momentum is an important factor. To increase movement in a positive direction requires force from many sources. Momentum is measured by the formula $\frac{1}{2}MV^2$. The mass consists of social and economic factors that can be added. Population, education, housing, food supplies, industrialization, recreation, wealth can be increased. The velocity is also important because it is squared. Momentum functions in society like a gyroscope or a flywheel. In a ship it will provide a counterforce for stability in rough seas. Inertial guidance systems of the space vehicles use the momentum of the earth for stability. If the velocity is increased too much, however, it can break up the system, and if it is slowed down too much the instability increases. Whenever the momentum is changed, the system feels the effect as either favorable or unfavorable.

For the past thirty years I have worked hard to increase the upward curve, and there are signs of improvement. State and federal governments looked at the great wealth of the cities and took tax dollars away, returning only a minimum for relief of the poor people who were living in them. Now some consideration is being given to correcting this serious imbalance. In Detroit itself private investment is also increasing. Financial institutions have built new headquarters in the central business district. Utility companies have added to their facilities. And a number of major industries, under the leadership of Henry Ford II, have pooled resources to build a new housing and commercial complex on the banks of the Detroit River. It is called Renaissance Center and everyone approves the promise of its name.

The Doxiadis study described the future Detroit as a region because transportation and communication technologies are already sufficiently developed to make it one. Even today, it is a region in every sense except for the governmental systems which lag far behind the needs of people. Businessmen are beginning to give more thought to general social conditions, including the organization of decision-making about them, and are working with political leaders to increase the rate of change. Regional authorities are being formed. A Southeastern Michigan Council of Governments has been developed as a voluntary association of local governments within a seven-county region. Still, the economic and technological changes that shape our daily urban experience are moving very rapidly and governmental response is relatively slow.

In a subsequent study, Doxiadis also discovered evidence that a still larger concentration of people is being formed in what he called the Great Lakes megalopolis. If past trends hold, a continuous built-up area will soon reach from Milwaukee and Chicago on the west, through Detroit, Toledo and Toronto in the center, to Cleveland and Pittsburgh on the east. It was the automobile and telephone and great quantities of energy that made the urban Detroit region possible, and perhaps inevitable. What is required to produce the Great Lakes megalopolis is the invention of somewhat more sophisticated technology and the discovery of still more energy. Both of these are well within the present capability of the human mind. Such a megalopolis is already in an advanced stage of development between Boston and Washington. But it will require a wholly new level of thinking to bring about the desirable management of public matters on this scale.

It is clear that the businessmen and engineers of tomorrow will have many new frontiers to explore as they seek to improve the living conditions of the children and grandchildren of the world. Personally, I view the prospect as stimulating.

CHAPTER VIII

A SUMMING UP

As anyone who has read this far will have observed, I am by nature an outward-going rather than an introspective man, and I am not given to self-analysis. I have worked with many who are introspective, and they have been very creative people who have provided their own insights and their own contributions to the world. I enjoy expressing myself in action, so this book is chiefly the reminiscences of my activities as a businessman.

From the beginning of my professional career I took it for granted that I had a responsibility to improve the company for which I worked. Everyone expects this of businessmen, even those who are critical of business in general. It is necessary to earn a profit because, whether we measure our activity under a free or a controlled economic system, in an industrial or emerging nation, we must have a standard by which to allocate the resources to increase our production. In a simple agricultural community, the farmer must save seed for the next planting. In a complicated industrial society, profitability is the source of improved benefits to customers, employees and shareholders.

Not everyone takes it for granted, however, that a businessman can also serve his country and his world. Although I worked for Public Service in New Jersey for nineteen years, it was not until I joined the Office of Production Management that my awareness of economic interdependence began to broaden sufficiently to show me how great the possibilities were. The experience I gained in the War Production Board was invaluable. It gave me some conception of what it takes nationally and internationally to bring together a whole spectrum of forces—people, industrial organizations, governmental structures, natural resources—in such a way as to meet our civilian and defense requirements. It was the largest organizational effort the country had ever undertaken and it united us as we had seldom been before on the nature of our national goals.

Because of my experiences in the mobilization and then in the war, I was able to participate in organizing first for peace and then for reconstruction. The critical need in rebuilding Europe and Asia was improved productivity, and my special part was to increase the supplies of electric energy to make this possible. My efforts brought me together with the most knowledgeable people in unions, management and government both in the United States and overseas. It was a privilege to be with them, and I count these associations among the riches of my life.

There is no substitute for able people who are determined to do something. I was determined to listen. If I really listened first of all, I could then add something from my background and we could work together. It is necessary to analyze problems, but I was never content with just being critical. I always moved to the constructive approach of how to make things better than they had been. I sought ways of avoiding confrontation and gaining cooperation. I just wanted to do a job well, to bring everything I had learned in the past to what I was engaged in. And I always found able men in both public and private enterprise, so I am convinced we will do best under a mixed system.

The affairs of mankind are always getting out of balance and we are kept busy restoring them. Experience has taught me a great deal about how to deal with the four imbalances with which I have been particularly concerned. Although all too often we are frustrated and sometimes turn to the terrible violence of war, most of the time we succeed in solving problems and in making measurable progress. In my lifetime the standard of living for millions of Americans and other people around the world has improved greatly. I have confidence that we can extend the material well-being of all people and that by the year 2000 everyone will enjoy a higher quality of life. Such is the power of the intellect that if mankind can conceive these things, we can accomplish then. I have seen it happen time and again. The greatest force on earth other than that of the Almighty is the human mind.

I am also confident that as we increase the quantity of energy available to do the world's work we free time for other than material activities. We free the spirit of people as well as their bodies. We are able to be generous, tolerant and concerned for others. I have observed this in the millions of hours that people put into volunteer work in communities and churches, in education and the arts, for youth and the elderly, at home and abroad. I believe there has been a measurable increase in such riches of the spirit as well as in material wealth. This is certainly not the best of all possible worlds but it is a great deal better than it was for citizens of industrial countries, and we know how to make it better for everyone. This has been my creed.

CHRONOLOGY

1897 - Born October 8, in Marietta, Ohio.

1922 - Graduated from Cornell University. Joined Public Service Electric and Gas Company of New Jersey as a cadet engineer.

1941 - Loaned to the Office of Production Management which became the War Production Board after the declaration of war.

1943 - Employed by The Detroit Edison Company. Commissioned a lieutenant colonel and sent to the Mediterranean Theater of Operations.

1944 - Appointed Chief of the Public Utilities Section of Supreme Headquarters Allied Expeditionary Forces (SHAEF).

1945 - Awarded Croix de Guerre with Palm and named Chevalier, Legion of Honor (France); Honorary Officer, Order of the British Empire; Commander, Order of Orange Nassau (The Netherlands). Returned to Detroit Edison.

1946 - Bronze Star and Legion of Merit (United States); Officer, Order of Leopold (Belgium).

1948 - Participated in the Marshall Plan in Europe. Established the semi-annual power surveys of the Edison Electric Institute and introduced similar surveys in Europe and Japan. Worked with AEC on atomic energy.

1951 - President and general manager of Detroit Edison.

1953 - Officer, Legion of Honor (France); Royal Order of the Phoenix (Greece). Congressional approval for interconnections between Detroit Edison and Hydro Electric Power Commission of Ontario, Canada.

1954 - President and chief executive officer of Detroit Edison.

1955 - First international conference on Peaceful Uses of Atomic Energy in Geneva, Switzerland. Broke ground for Enrico Fermi Power Plant #1 with a sodium-cooled fast breeder reactor.

1956 - Commander, Order of Merit (Italy). Participated in the Hartley report on *Europe's Growing Needs of Energy - How Can They Be Met?*

1958 - Order of Merit (West Germany).

1960 - Grand Medal of Honor (Austria); Commander, Order of Leopold (Belgium). Participated in the Robinson report, *Towards a New Energy Pattern in Europe.*

1961 - Order of the Rising Sun, Third Class (Japan). Energy study of Tunisia.

1962 - Grand Master, Order of the White Rose (Finland).

1963 - Fermi #1 achieved self-sustaining nuclear operation. Power survey of Taiwan.

1964 - Chairman of the board and chief executive officer of Detroit Edison. Commander, Legion of Honor (France). Power survey of Korea.

1965 - Commander, Royal Order of the Phoenix (Greece). Energy study of India and power survey of the Philippines. Began the Developing Urban Detroit Area Research Project.

1966 - Grand Officer, Order of St. Alexander (Bulgaria); Commander, Royal Order of Vasa (Sweden). Power survey of Thailand.

1967 - Order of Homayoun, Third Class (Iran); Grand Officer, Order of Merit (Italy). Power survey of Iran. *Energy and the Michigan Economy.*

1968 - Chairman, International Executive Council of the World Energy Conference.

1970 - Order of Industrial Service (Korea).

1971 - Public Works Medal, First Order (South Vietnam). Power survey of the Republic of Vietnam. Chairman of the board, Detroit Edison.

1972 - President, Overseas Advisory Associates, Inc.

1974 - Chairman, World Energy Conference, Detroit.

1975 - Retired as chairman of the board, The Detroit Edison Company.